The Cost of Environmental Degradation

The Cost of Environmental Degradation

Case Studies from the Middle East and North Africa

Lelia Croitoru and Maria Sarraf
Editors

THE WORLD BANK
Washington, D.C.

ISBN: 978-0-8213-8318-6
eISBN: 978-0-8213-8396-4
DOI: 10.1596/978-0-8213-8318-6

Library of Congress Cataloging-in-Publication Data
Croitoru, Lelia.
  The cost of environmental degradation : case studies from the Middle East and North Africa / Lelia Croitoru and Maria Sarraf.
    p. cm.
  ISBN 978-0-8213-8318-6 — ISBN 978-0-8213-8396-4 (electronic)
  1. Environmental degradation—Middle East—Case studies. 2. Environmental degradation—Africa, North—Case studies. 3. Middle East—Environmental conditions—Case studies. 4. Africa, North—Environmental conditions—Case studies. I. Sarraf, Maria. II. Title.
  GE160.M628C76 2010
  363.700956—dc22

                                                                                    2010010933

Cover photo: Corbis
Cover design: Naylor Design, Washington, DC

Contents

Figures

Tables

Photos

Foreword

Environmental degradation is costly—to individuals, to societies, and to the environment. This book, edited by Lelia Croitoru and Maria Sarraf, makes these costs clear by examining a number of studies carried out over the past few years by the World Bank's Middle East and North Africa Region. Even more important than estimating the monetary cost of environmental degradation (COED), however, are the clear guidance and policy implications derived from these findings.

The Middle East and North Africa region is large—stretching from Morocco in the west to the Islamic Republic of Iran in the east—and is home to more than 310 million people in 13 countries. Most of these countries are heavily urban, fairly arid, and resource-poor (except for those countries with substantial reserves of oil and gas). Although annual per capita incomes in 2009 ranged from a low of US$1,090 in Djibouti to more than US$9,000 in Libya, the region's average (US$2,820 per capita) is close to the World Bank's worldwide average for middle-income countries (US$3,053 per capita). Nevertheless, the COEDs in the country studies reported in this book range from 2.1 percent to 7.4 percent of each country's gross domestic product. These are major, potentially avoidable costs in countries that tend to have high population growth rates and struggle to increase their average per capita incomes.

This volume presents a new approach to estimating the impacts of environmental degradation. In the past, when government officials asked researchers the simple question "How large are the impacts of environmental degradation?" the response was often an emphatic "Large!"—a rather imprecise number. The strength of this work is that it actually quantifies in economic terms how large is "large" and thereby gains the attention of decision makers and offers specific insights for improved policy making.

Using a range of available valuation techniques and country-specific data (based on a 25-year time horizon and a 4 percent discount rate), the studies present annual values for the COEDs in several Middle Eastern and North African countries. The book focuses on major environmental challenges to the quality of air, agricultural lands, water, forests, and coastal areas. Each environmental issue is illustrated by a chapter-length case study. Given recent events—the 2010 oil spill in the Gulf of Mexico—the chapter on estimating the COED resulting from oil spill and demolition waste is especially timely.

Croitoru and Sarraf make an important contribution to the literature by making this work available. Although they hasten to note that the results do not lend themselves to strict country-to-country comparisons (because of inherent data-related and other constraints), the presentation of results on an annual basis, and by major environmental damage category, does create an interesting snapshot of environmental damage costs in the Middle East and North Africa region. In addition, each chapter suggests appropriate policy responses that differ depending on the root cause of the problem—such as unsustainable resource use and management (for example, in forestry, agricultural land, and coastal areas) or mishandling of the pollution byproducts of otherwise desirable economic activity (as in the cases of air and water pollution or waste disposal).

Finally, this book demonstrates the benefits of doing a coordinated, regional COED analysis that builds on the country-level studies. This two-tiered approach produces important synergies—in terms of both the methodologies used and the lessons learned. The book is definitely a case in which "the whole is greater than the sum of the parts." I hope this work will inspire others to conduct similar regional analyses.

John A. Dixon
Lead Environmental Economist (retired)
The World Bank

Acknowledgments

The World Bank's Cost of Environmental Degradation (COED) studies were initiated under the Mediterranean Environmental Technical Assistance Program (METAP). METAP is a partnership between countries of the Mediterranean region and multilateral donors aimed at strengthening the structure of environmental management, formulating environmental policies, and developing environmental projects in the Middle East and North Africa region. This book owes much to Sherif Arif, the former METAP coordinator.

Sustainable MED is a regional initiative to integrate the environment within the economic development agenda of the Mediterranean following a shared vision. The METAP and the Sustainable MED programs provided financial support for this book.

The chapters of this book draw heavily on the original COED reports, although they have been extensively rewritten, and in many cases, the estimates have been revised or updated with newly available data. The reports benefited greatly from the assistance of many governmental institutions and other stakeholders in the countries being studied. Special thanks go to the Ministries of Environment of the Islamic Republic of Iran, Jordan, Lebanon, Morocco, and Tunisia for their contributions to this innovative effort.

The team would like to thank Luis Constantino, Julian Lampietti, Dahlia Lotayef, Stefano Pagiola, and Thor Sigvaldason for their continuous support and advice throughout the preparation of this book. Richard Crabbe, Mary Fisk, Stuart Tucker, and the team at the World Bank Office of the Publisher worked hard to ensure that the book is published in a timely fashion and reaches the widest possible audience.

Abbreviations and Symbols

AFD	Agence Française de Développement
ANPE	Agence Nationale de Protection de l'Environnement (Tunisia)
CAS	Country Assistance Strategy
CDM	Clean Development Mechanism
CEA	Country Environmental Analysis
CEDRE	Centre de Documentation, de Recherche et d'Expérimentation sur les pollutions accidentelles des eaux
CEP	Caspian Environment Programme
COED	Cost of Environmental Degradation
COI	Cost of Illness approach
CNSS	Caisse Nationale de Securité Sociale (Tunisia)
CV	contingent valuation
DALYs	disability-adjusted life years
DGACTA	Direction Générale de l'Aménagement et de Conservation des Terres Agricoles (Tunisia)
DGETH	Direction Générale des Grands Travaux Hydrauliques (Tunisia)
DGRE	Direction Générale des Ressources en Eau (Tunisia)
DNAPLs	dense non-aqueous phase liquids

DPL	Development Policy Loan
DSSB	Direction des Soins de Santé de Base (Tunisia)
EC	European Commission
EER	Energy Environment Review
FAO	Food and Agriculture Organization of the United Nations
FODEP	Fonds de Dépollution (Tunisia)
FRWO	Forest, Range and Watershed Organization (Islamic Republic of Iran)
FU	forage unit
GAC	Government Appointment Committee (Lebanon)
GAM	Greater Amman Municipality (Jordan)
GDP	gross domestic product
GEF	Global Environment Facility
GIS	Geographical Information Systems
GLASOD	Global Assessment of Soil Degradation
GNI	gross national income
GOL	Government of Lebanon
HCA	human capital approach
ICM	Integrated Coastal Management
IFAD	International Fund for Agricultural Development
IFO	intermediate fuel oil
INS	Institut National de la Statistique (Tunisia)
INSTM	Institut National des Sciences et Technologies de la Mer (Tunisia)
IPPS	Industrial Pollution Projection System
IUCN	International Union for Conservation of Nature
MAERH	Ministère de l'Agriculture, de l'Environnement et des Ressources Hydrauliques (Tunisia)
MADREF	Ministère de l'Agriculture, du Développement Rural, des Eaux et Forêts [Ministry of Agriculture, Rural Development, Waters and Forests] (Morocco)
MAI	mean annual increment
MARH	Ministère de l'Agriculture et des Ressources Hydrauliques (Tunisia)
MAMVA	Ministère de l'Agriculture et de la Mise en Valeur Agricole (Morocco)
MATUHE	Ministère de l'Aménagement du Territoire, de l'Urbanisme, de l'Habitat et de l'Environnement (Morocco)
MDBs	Multilateral Development Banks

MEAT   Ministère de l'Environnement et de l'Aménagement du
            Territoire (Tunisia)
MEDD   Ministère de l'Environnement et du Développement
            Durable (Tunisia)
METAP   Mediterranean Environmental Technical Assistance
            Program
MACCSL   Mine Action Coordination Center South Lebanon
MOA   Ministry of Agriculture (Lebanon)
MOJA   Ministry of Jihad and Agriculture (Islamic Republic
            of Iran)
MSW   municipal solid waste
NEAP   National Environmental Action Plan
NDO   National Demining Office (Lebanon)
NBSAP   National Biodiversity Strategy and Action Plan (Islamic
            Republic of Iran)
NGO   nongovernmental organization
NPV   net present value
NWFPs   nonwood forest products
ONAS   Office National de l'Assainissement (Tunisia)
ONTT   Office National du Tourisme Tunisien (Tunisia)
OTED   Observatoire Tunisien de l'Environnement et du
            Développement Durable (Tunisia)
PISEAU   Projet d'Investissement dans le Secteur de l'Eau (Tunisia)
PCM   Presidency of the Council of Ministers (Lebanon)
RR   relative risk
RSS   Royal Scientific Society
SCET   Société Centrale pour l'Equipement du Territoire
            (Tunisia)
SCI   Statistical Center of (the Islamic Republic of) Iran
SDC   Swiss Agency for Development and Cooperation
SONEDE   Société Nationale d'Exploitation et de Distribution des
            Eaux (Tunisia)
TEV   total economic value
TSP   total suspended particulates
UNDP   United Nations Development Programme
UNEP   United Nations Environment Programme
UNESCO   United Nations Educational, Scientific and Cultural
            Organization
UNFCCC   United Nations Convention on Climate Change
UNICEF   United Nations Children's Fund

UXOs	unexploded ordnances
VSL	value of a statistical life
WHO	World Health Organization
WTP	willingness to pay
WWF	World Wildlife Fund

Symbols

%	percent
°C	Celsius degrees
C	carbon
CO	carbon monoxide
CO_2	carbon dioxide
du	dunum (1 dunum = 1,000 square meters)
g	gram(s)
ha	hectare(s) (1 hectare = 10,000 square meters)
HC	hydrocarbon
km	kilometer(s) (1 kilometer = 1,000 meters)
q	quintal(s) (1 quintal = 100 kilograms)
$\mu g/m^3$	micrograms per cubic meter
m	meter
m^2	square meter
m^3	cubic meter
m^3/ha	cubic meter per hectare
NO_x	nitrogen oxides
SO_2	sulphur dioxide
t	ton
t/ha	ton per hectare
tCO_2e	tons carbon dioxide equivalent
US$	United States dollars
cSt	CentiStokes
H_2S	hydrogen sulphide
Pb	lead
PM	particulate matter
$PM_{2.5}$	particulate matter with a diameter smaller than 2.5 microns
PM_{10}	particulate matter with a diameter smaller than 10 microns
n.a.	not applicable
..	negligible
no.	number

CHAPTER 1

Introduction

Lelia Croitoru and Maria Sarraf

That environmental degradation can be a by-product of economic activities is no secret. Industrial production often discharges pollutants into clean rivers and air, preventing the use of these resources for other purposes and harming the health of those exposed to them. Unsustainable agricultural practices can reduce crop productivity and cause dam sedimentation. Overexploitation of groundwater increases pumping costs and, if it leads to saltwater intrusion, may make aquifers unusable.

These and many other forms of environmental degradation cause real costs to the economy and to people's welfare. Yet these costs often go unmeasured, and thus, their magnitude is largely unknown. Therefore, a country typically has insufficient information about the level of environmental damage, let alone about the way to reduce or reverse the damage.

Until recently, most available studies have estimated the costs of environmental degradation for specific sites or industries. Since 2000, however, the World Bank has conducted a systematic effort to measure the cost of environmental degradation (COED) at the national and local levels in several countries of the Middle East and North Africa region.[1] This book brings together the best COED case studies completed in several of the region's countries and summarizes the case studies' impacts on the ground at national and regional levels.

In addition to covering several countries, the book stands out by applying updated methods and techniques to estimate the environmental damages caused by a conflict.

Measuring the Cost of Environmental Degradation

The collection of case studies in this book updates the original COED work published in stand-alone World Bank reports. *It should be noted that these case studies do not necessarily reflect the highest environmental priorities in each country.*[2] Rather, the case studies represent the application of the most suitable methods and techniques, in the context of the available data, to estimate the costs of several types of environmental damage.

The objective of the COED studies is to estimate the *annual* cost of environmental degradation. The studies measure in monetary terms the present and future impacts caused by the environmental damage occurring in one given year, or the year of reference. Each final estimate is then stated in terms of *percentage of the country's gross domestic product (GDP)* for the year of reference. This estimate expresses the COED in terms comparable to the most widely used economic indicators, making it immediately intelligible to both policy makers and the general population (Sarraf 2004). It also allows for comparisons of (a) COED estimates among different countries and (b) the costs of different types of environmental damage within the same country.

The COED usually measures the damage caused to several environmental categories: water, air quality, agricultural land, forests, waste, and coastal zone. Spatially, the analysis can be done at the local level (for example, a city or a coast), the national level (a country), the multinational level (several countries), or even the regional level (for example, the entire Middle East and North Africa region). Depending on the objective of the analysis, it can focus on environmental categories overall or on just one or a few categories.

The environmental damage categories can be further summarized as damages to the following:

- *Agricultural land:* losses of agricultural productivity on croplands and rangelands due to unsustainable practices
- *Forests:* losses of forest goods (for example, timber, firewood, and non-wood forest products) and services (such as watershed protection and recreation) due to deforestation and forest degradation

- *Water:* impacts on major economic sectors of water salinity,[3] contamination, waterlogging, dam sedimentation, and overexploitation of groundwater
- *Air quality:* impacts of air pollution on health (costs of mortality and morbidity from airborne diseases) and the environment (through reduced visibility and aesthetic value of landscape)
- *Waste:* impacts on the environment and public welfare of inappropriate waste collection, transport, and disposal
- *Coastal zone:* losses of recreational and landscape value due to unsustainable coastal activities.

The COED estimates are based on standard valuation techniques (Dixon and others 1994), with the specific approaches chosen largely based on data availability in each country. When sufficient data are available, losses are assessed using demand curve approaches, ranging from direct methods based on market pricing to indirect ones such as the substitute goods method and production function. When the data are insufficient to apply demand curve approaches, losses are assessed using cost-based methods or, whenever meaningful, the benefit transfer method by applying results from other studies undertaken in similar contexts. In still other cases, relevant information is not available at all. Thus, certain benefits could not be estimated. Chapters 2 through 6 summarize the methods and approaches for estimating the impacts in each environmental category. For more detailed descriptions of each method, see Bolt, Ruta, and Sarraf (2005).

To make the estimates as comparable as possible across countries, all country studies use common baseline assumptions. For example, estimating the impacts over time caused by the degradation occurring today is based on a time horizon of 25 years and a discount rate of 4 percent (World Bank 2005). In addition to these assumptions, each chapter of the book presents in detail the individual assumptions used for the valuation.

Despite the effort to ensure a common and comprehensive framework for valuation in all countries, the estimation efforts unavoidably encountered some limitations. First, the base year of analysis differs across country studies, albeit within a narrow range of years. Second, the same damage is sometimes estimated by applying different methods across countries, depending on the available data in each country. As a result, the overall COED and the cost of each environmental category are not directly and accurately comparable across countries. Third, data-related constraints made it difficult for the country studies to capture the entire range of damages. Therefore, the final results represent *orders of magnitude*, which generally *underestimate* the true value of damage and should not be interpreted as accurate estimates.

The Cost of Environmental Degradation in the Middle East and North Africa

The six case studies are drawn from a region that extends from Morocco in the west to the Republic of Yemen in the east and includes the 13 countries listed in table 1.1. The region is home to about 313 million people, or 5 percent of the world's population. Its average income of US$2,820 per capita is close to the average income for middle-income countries of US$3,053, but with wide variations. Yemen's income level is slightly below

Table 1.1   Comparative Data on Middle East and North African Economies

Economy	Population (millions)	GNI (US$/capita)	Agricultural land (% of land area)	Forests (% of land area)	Internal freshwater resources (m³/capita)	Urban population (% of total)
Algeria	34.0	3,620	17	1.0	332	63.9
Djibouti	0.8	1,090	73	0.2	360	86.5
Egypt, Arab Rep.	75.0	1,580	4	0.1	24	43.0
Iran, Islamic Rep.	71.0	3,540	29	6.8	1,809	67.4
Iraq	—	—	23	1.9	—	—
Jordan	5.7	2,840	11	0.9	119	82.6
Lebanon	4.1	5,800	38	13.3	1,172	86.7
Libya	6.2	9,010	9	0.1	97	85.1
Morocco	31.0	2,290	68	9.8	940	59.3
Syrian Arab Rep.	20.0	1,780	76	2.5	352	50.8
Tunisia	10.0	3,210	63	6.8	410	65.7
West Bank and Gaza	3.7	1,290	62	1.5	—	71.7
Yemen, Rep.	22.0	870	34	1.0	94	27.7
Middle East and North Africa	**313.0**	**2,820**	**22**	**2.4**	**728**	**57.5**
Middle-income countries	3,088.0	3,053	35	33.8	8,754	54.5
World	6,538.0	7,448	38	30.5	6,778	49.1

Source: World Bank (2009).
Note: GNI = gross national income, m³ = cubic meters, — = not available.

that of low-income countries, while Libya's nearly reaches the upper level of the high middle-income countries.[4]

The Middle East and North Africa, as a region, is facing important threats to natural resources that are already scarce. Forests cover just 2.4 percent of the region's land, of which more than 75 percent is in the Islamic Republic of Iran and Morocco (World Bank 2009).[5] With an average freshwater availability of 728 cubic meters per capita, many countries cannot meet the current water demand. The extent of agricultural land also varies across countries, most of which depend on food imports. In many of the region's countries, high proportions of their populations live in urban areas, especially in the coastal zones.

Population growth will contribute to a decrease in already-scarce water availability of 50 percent by 2050, and climate change is predicted to bring more frequent and severe droughts and floods (World Bank 2007). In many areas, water overuse and misuse are further degrading water supplies and threatening the population's welfare. Inappropriate irrigation practices may increase water salinity and reduce agricultural productivity. Inadequate access to clean water and sanitation results in waterborne diseases, particularly among children. Dam sedimentation and groundwater overexploitation threaten future water availability for drinking and irrigation.

Chapter 2 examines these issues in detail in the case of Tunisia and estimates the cost of water degradation at *0.6 percent of the country's GDP* in 2004. Although direct comparisons are difficult, the cost of water degradation is higher in other Middle Eastern and North African countries. The Islamic Republic of Iran, for example, stands out with a cost of water degradation as high as 2.8 percent of GDP in 2000 (World Bank 2005). In the other countries in the region, the total cost associated with water degradation, inadequate water supply, and poor sanitation ranges from 0.8 percent of GDP in Algeria and Jordan to 1.2 percent of GDP in Morocco.

Transport, power generation, and industry account for the bulk of air-pollutant emissions in several Middle Eastern and North African countries. Air pollution is a growing problem, particularly in heavily populated cities such as Cairo in the Arab Republic of Egypt and Tehran in the Islamic Republic of Iran. Long-term exposure to combustion-related fine particulate air pollution is an important environmental risk factor for lung cancer and cardiopulmonary mortality (Pope and others 2002). In addition to causing health problems, air pollution also deteriorates the environment by reducing visibility and the landscape's aesthetic value.

Chapter 3 addresses these issues in detail in the case of Jordan, where the cost of air pollution is estimated at *1.2 percent of GDP* in 2006. Here, air pollution affects selected road-traffic and industrial hot spots without affecting tourism poles or overall air quality at the national level. The cost may be a good deal higher in other countries such as Egypt (2.1 percent of GDP) and the Islamic Republic of Iran (1.6 percent of GDP) because of the impacts of air pollution on deaths and illnesses in those countries' most populated cities (World Bank 2002, 2005).

Forests in the Middle East and North Africa cover about 20.6 million hectares, or just 0.5 percent of the world's forests (World Bank 2009). Despite their scarcity in the region compared with other regions, these forests provide a wealth of valuable goods and services, such as watershed protection, biodiversity, and other nonwood forest products. Forests are threatened by deforestation, however, which often induces a complete or partial loss of the goods and services on which forest communities highly depend.

Chapter 4 addresses these issues in the case of the Islamic Republic of Iran, the most-forested country of the region, and estimates the cost of deforestation and forest degradation at *0.7 percent of GDP* in 2002. More than 80 percent of this value occurs in the deforested areas, where losses of forest services (for example, watershed protection, recreation, and carbon sequestration) are more significant than the losses of timber and other extractive values.

Most Middle Eastern and North African countries import at least 50 percent of the calories they consume.[6] Growing populations and increased demand for agricultural products such as cereals may trigger even more reliance on imports in the future. Although many countries in the region stress the importance of maintaining or increasing agricultural productivity, inappropriate cropping or irrigation practices and overgrazing of rangeland contribute to degradation of agricultural land and reduce productivity over time.

Chapter 5 analyzes these issues in detail in the case of Morocco and estimates the cost of agricultural land degradation at *0.4 percent of GDP* in 2000. Cropland degradation from unsustainable practices accounts for most of this cost. The cost of agricultural land degradation is comparatively higher in the other Middle Eastern and North African countries, peaking in the Islamic Republic of Iran (1.7 percent of GDP) mostly because of the impact of soil salinity on crop productivity.

The Middle East and North Africa region is also, unfortunately, affected by conflicts that have caused enormous losses in terms of deaths and injuries. Such violent conflicts also damage the environment in ways that are often expensive to restore, if not irreversible.

Chapter 6 provides a comprehensive valuation of the environmental damages caused by the 34-day conflict in Lebanon in July and August 2006. The hostilities killed almost 1,200 people, injured more than 4,400, and severely damaged the country's infrastructure. Moreover, the bombing of a power plant in Jiyeh caused the spill of about 12,000 to 15,000 tons of oil into the Mediterranean Sea. The conflict left enormous amounts of demolition and military waste, including debris, rubble, and unexploded ordnances (UXOs).

This chapter estimates the cost of environmental damage associated with the oil spill and waste at *2.4 percent of GDP* in 2006. The impacts of demolition and military waste account for 1.4 percent of GDP, primarily because of the high costs of disposing of demolition waste, of casualties, and of agricultural losses caused by UXOs in South Lebanon. The damage due to the oil spill represents 1 percent of GDP, mainly because of the high costs of cleaning up the oiled waste, the cost of oil burned and spilled, and the income losses from coastal services (for example, hotels and restaurants).

The COED work conducted in the Middle East and North Africa has had significant impact on the ground. It has not only raised the countries' awareness about the magnitude of the existing environmental damage, but also become the driving force behind concrete actions to incorporate the COED into the investment and policy-making processes.

Chapter 7 summarizes the COED's main impacts in the Middle East and North Africa in terms of strengthening the policy dialogue, capacity building, institutions, and environmental investments.

The estimates of damage to each environmental category and natural resource in selected Middle Eastern and North African countries are part of a wider effort of valuing the full COED in these countries. Figure 1.1 summarizes the main results for selected countries in the region, including each country case study in this book. Although the estimates are not directly comparable,[7] figure 1.1 shows that the COED ranges from as little as 2.1 percent to as much as 7.4 percent of GDP for different countries and years. Once again, these costs are order-of-magnitude estimates and should be interpreted as such.

The COED is now a product increasingly demanded by the countries' governments and a flagship of the World Bank's Middle East and North Africa Region. At the methodological level, this book provides the best case studies from select Middle Eastern and North African countries as

Figure 1.1   Estimated COED in Selected Middle Eastern and North African Countries

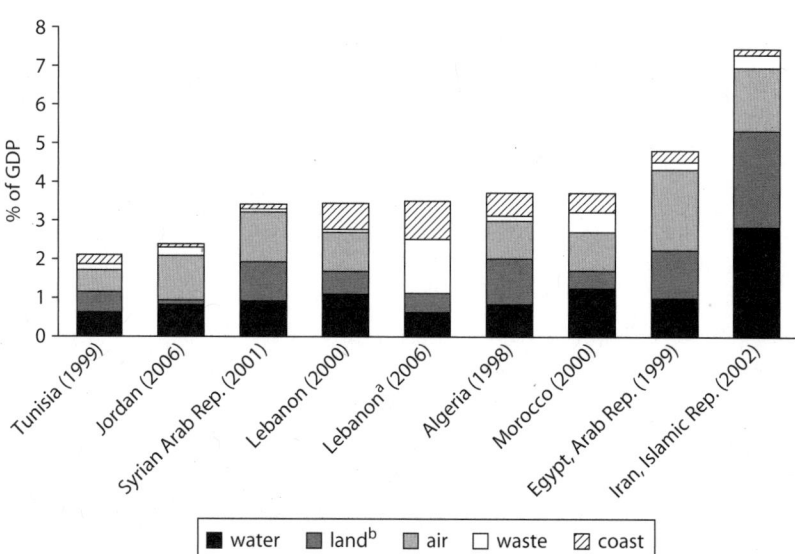

Source: Authors' calculations.

Note: Horizontal axis titles are countries studied and their respective years of reference for valuation, in order of increasing COED as a percentage of GDP. Keys designate environmental damage categories, shown as a proportion of the COED.

a. The 2006 COED study of Lebanon focuses on the July–August 2006 conflict. Damage to the coastal zone refers to oil spill damages. Damage associated with the waste sector refers to demolition, military, and medical waste. Unlike other estimates, which are annual, this estimate reflects the damage to the environment from a 34-day conflict.

b. Costs attributed to "land" include damages to both agricultural land and forests.

well as a methodology that can be applied in other parts of the world. At the policy level, the book offers a holistic picture of the estimated values at national and regional levels and sheds light on the concrete implications for policy making in several Middle Eastern and North African countries.

Notes

Lelia Croitoru is an Environmental Economist at the World Bank, Middle East and North Africa Region, 1818 H Street, NW, Washington, DC 20433, USA.

Maria Sarraf is a Senior Environmental Economist at the World Bank, South Asia Region, MC10-1019, 1818 H Street, NW, Washington, DC 20433, USA.

1. The countries include Algeria, the Arab Republic of Egypt, the Islamic Republic of Iran, Jordan, Lebanon, Morocco, the Syrian Arab Republic, and Tunisia.

2. For example, estimating the cost of water degradation in Tunisia (Chapter 2) does not mean that water degradation is the worst environmental problem in Tunisia or elsewhere in the Middle East and North Africa. Similarly, valuing the air pollution in Jordan (Chapter 3) should not imply that air pollution is the highest environmental priority in Jordan or in the Middle East and North Africa region.

3. The focus is not on natural water salinity but on increased salinity caused by inappropriate irrigation practices.

4. The World Bank (2009) defines low-income economies as those with a gross national income (GNI) per capita of US$935 or less in 2007. Middle-income economies are those with a GNI per capita between US$935 and US$11,456.

5. Forests cover 45 percent of land area in the Latin America and Caribbean region and 28 percent of land area in the East Asia and Pacific region. The world average is 30 percent.

6. Food consumption statistics are from the FAOSTAT database of the Food and Agricultural Organization of the United Nations (FAO). http://faostat.fao.org.

7. Directly comparable estimates are unavailable because of the different years of reference and sometimes different valuation methods used.

References

Bolt, K., G. Ruta, and M. Sarraf. 2005. *Estimating the Cost of Environmental Degradation: A Training Manual in English, French, and Arabic.* Environment Department Papers. Washington, DC: World Bank.

Dixon, J., L. F. Scura, R. A. Carpenter, and P. B. Sherman, eds. 1994. *Economic Analysis of Environmental Impacts.* London: Earthscan.

Pope, C. A., R. T. Burnett, M. J. Thun, E. E. Calle, D. Krewski, K. Ito, and G. Thurston. 2002. "Lung Cancer, Cardiopulmonary Mortality, and Long-Term Exposure to Fine Particulate Air Pollution." *Journal of the American Medical Association* 287 (9): 1132–41.

Sarraf, M. 2004. "Assessing the Costs of Environmental Degradation in the Middle East and North Africa Region." Environment Strategy Notes 9, World Bank, Washington, DC.

World Bank. 2002. *Arab Republic of Egypt: Cost Assessment of Environmental Degradation.* Sector Note, Report 25175-EGT. Rural Development, Water and Environment Department; Middle East and North Africa Region. Washington, DC: World Bank.

———. 2005. *Islamic Republic of Iran: Cost Assessment of Environmental Degradation.* Sector Note, Report 32043-IR. Middle East and North Africa Region. Washington, DC: World Bank.

———. 2007. *Making the Most of Scarcity: Accountability for Better Water Management Results in the Middle East and North Africa*. Directions in Development Series. Washington, DC: World Bank.

———. 2009. *The Little Green Data Book 2009*. Washington, DC: World Bank.

CHAPTER 2

Water Degradation: The Case of Tunisia

Lelia Croitoru, Maria Sarraf, Fadhel Ghariani, Mohamed Salah Matoussi, and Hamed Daly-Hassen

With water availability of 470 cubic meters per capita—or less than 50 percent of the Middle East and North Africa region's average—Tunisia is already a water-stressed country (OTED 2004; MARH 2005e). The limited water supply is unevenly distributed across the country. High demand for water resources has intensified their use, creating serious challenges such as increasing degradation and risk of depletion. This chapter estimates the annual cost of water degradation in Tunisia by addressing the effects of water salinity,[1] contamination, waterlogging, dam sedimentation, and overexploitation of groundwater on the major economic sectors. The chapter follows up on an earlier study that estimated the annual cost of environmental degradation of water, air, land, coastal zones, and waste in Tunisia (Sarraf, Larsen, and Owaygen 2004).

Overview of the Water Sector

Water resources in Tunisia are scarce. Potential conventional water resources are estimated at 4.6 billion cubic meters per year, of which 54 percent are surface water and the rest groundwater. About 1.8 billion cubic meters of surface water and 1.9 billion cubic meters of groundwater are mobilized annually. Agriculture is by far the largest water consumer. In 2004, 2.1 billion cubic meters were allocated to agriculture, or approximately 83 percent of the total water demand (Lebdi 2005). Drinking water and domestic tourism consumed 318 million cubic meters during the same year, while the industrial sector consumed 122 million cubic meters (MARH 2005e).

Water is subject to high seasonal and interannual variability, which is particularly marked in the south. Water is also unevenly distributed geographically and does not always match the location of the demand. The interior regions (northwest and central west) of Tunisia provide most of the water resources, although the demand is relatively concentrated on the eastern seashore. This imbalance called for expensive water transfer works.

The main water-related problems in Tunisia are the following:

- *Scarcity*. Steadily increasing demand for water will increase water scarcity (World Bank 2004), which in turn will lead to greater reliance on unconventional resources such as desalination and waste-water treatment. Deterioration of water quality will also increase scarcity and, conversely, scarcity may cause degradation because of overexploitation.
- *Salinity*. Fifty-three percent of mobilized resources have salinity higher than 1.5 grams per liter (ITES 2002). In coastal areas, salinity can exacerbate scarcity; conversely, the scarcity may increase water degradation.
- *Chemical and bacteriological contamination*. This type of degradation affects both inland and coastal marine waters. Bacteriological contamination affects the networks supplying drinking water in rural areas because of lack of chlorination (World Bank 2004). Nationally, there are more than 750 sources of water pollution, discharging 155 million cubic meters of waste annually, which represents a source of possible contamination to both groundwater and surface water (MEDD 2007).
- *Dam sedimentation*. Silting reached 17 percent of the initial dam capacity in 2002 and is expected to fill a high share of the Sidi Salem dam's capacity of by 2020 (MAERH 2003).

Water scarcity, salinity, contamination, and dam sedimentation contribute to water degradation in Tunisia in the following ways:

- *Deteriorating water quality*. Whether caused by chemical or bacteriological contamination or by increased salinity due to intrusion of coastal and *chott* waters,[2] water degradation can affect all uses. For example, poor water quality for irrigation may reduce agricultural productivity in irrigated areas.

- *Reducing water quantity*. This occurs when the quantity of water extraction exceeds the sustainable level of exploitation.[3] For example, the overexploitation of groundwater reduces the groundwater level, which deepens the water table and may induce groundwater salination.

It should be noted that water resources have been subject to detailed planning for more than three decades (through the Master Plans of Water in Northern, Central, and Southern Tunisia), and many strategic studies have addressed the main sector issues, namely (a) integrated management and conservation of water resources, (b) economic efficiency of water use for irrigation, and (c) institutional reform and capacity strengthening in the water sector. At this stage, the government aims to transform the policy of water mobilization into a policy of demand management through technical, legal, economic, and institutional instruments.

Methodology

Water degradation can affect the total economic value (TEV) of a water ecosystem, including (a) direct uses, such as irrigated agriculture, fisheries, health, and tourism; (b) indirect uses, such as the decline in prices of urban land close to a polluted site; and (c) non-uses, such as biodiversity. In addition, water scarcity can induce groundwater overexploitation, which may lower the groundwater level, affecting all water uses. Table 2.1 summarizes the valuation methods used in this chapter to estimate the impacts of water degradation on different sectors. For a more comprehensive description of these methods, see World Bank 2007.

The selected valuation methods vary according to the type of impact, and they depend on the availability of information, as described below.

- *Impact on irrigated agriculture*. Estimating the impacts on agriculture of different types of water pollution is based on the "change in productivity" method. For example, the impacts of salinity, waterlogging, and insufficiently treated wastewater are valued through the reduction in irrigated productivity caused by a decline in water quality. The impact of dam sedimentation is estimated through both the "replacement cost" method (the additional cost required to build new reservoir capacity) and the "change in productivity" method (the losses in irrigated productivity because of reduced water availability).

Table 2.1   Estimated Impacts of Water Degradation and Valuation Methods

Type of value affected	Name of impact	Valuation method
Direct use value	**Impact on irrigated agriculture**	
	1. Salinity and waterlogging	1. Change in productivity
	2. Insufficiently treated wastewater	2. Change in productivity
	3. Dam sedimentation	3. Replacement cost, change in productivity
	Impact on fishing	Change in productivity
	Impact on health	
	1. Diarrhea	1. DALYs (mortality, morbidity), treatment cost, opportunity cost of time
	2. Typhoid fever	2. DALYs (morbidity), cost of treatment
	3. Hepatitis	3. DALYs (morbidity)
	Impact on tourist value	Additional travel cost
Non-use value	**Impact on biodiversity** (wetland degradation)	Replacement cost
All TEV components	**Groundwater overexploitation**	Replacement cost

Source: Authors.
Notes: This chapter does not estimate the indirect-use values due to the lack of appropriate data. DALYs = disability-adjusted life years.

- ***Impact on fisheries***. Similarly, the impact on fisheries is estimated through the "change in productivity" method—comparing the fish production in polluted and nonpolluted coastal areas and attributing part of the difference to water degradation.

- ***Impact on health***. We estimate the health-related effects of bacterial degradation of water on the extent of waterborne diseases in the population (diarrhea, hepatitis, and typhoid fever). The social costs due to mortality and morbidity are evaluated based on disability-adjusted life years (DALYs) (Mathers and others 2004; WHO 2006).[4] In addition, we consider the direct costs of treatment and care associated with these diseases and the opportunity cost of time spent with sick children.

- ***Impact on tourism***. Estimating the loss of tourist value in a polluted coastal area is based on the additional costs paid to travel to other, cleaner coastal areas. The transportation and subsistence costs that visitors actually pay are assumed to be a conservative proxy for the willingness to pay (WTP) for the improved tourist value of less-polluted coastal areas.

- *Impact on biodiversity*. We found no studies using stated preference methods (for example, the contingent valuation method) to estimate non-use values of aquatic ecosystems in Tunisia. Thus, we evaluate the loss of biodiversity based on the annual costs to replace (restore) environmental wetlands.

- *Groundwater overexploitation*. The cost of overexploiting groundwater is estimated based on the additional cost of pumping water from a deeper level and the cost of building replacement wells.

Water degradation causes damages both during the year when the degradation initially occurs and during the following years. For example, dam sedimentation in 2004 reduced water availability not only in 2004 but also in successive years. This chapter estimates the present value of the present and future annual flows of damages due to water degradation that occurred in 2004. The valuation uses a discount rate of 4 percent over a time horizon of 25 years.[5]

Cost of Water Degradation

The following subsections estimate the impact of water degradation on agriculture, fishing, health, tourism, and biodiversity, as well as the effect of groundwater overexploitation.

Agriculture

Agriculture is the largest water user in Tunisia. Salinity, waterlogging, and contamination of water often reduce agricultural productivity. In addition, reservoir sedimentation causes additional costs to replace dams or build new storage capacity.

Impact of salinity and waterlogging. Salinity and waterlogging reduce agricultural productivity in irrigated areas. Mhiri and Bousnina (1999) analyze the impacts of these factors on agricultural productivity in seven regions of the country: Kroumirie-Mogods, Central Tell, Lower Tell, Dorsale, Western Dorsale, Centre, and South. They quantify the loss of agricultural productivity in each region according to pedology, rainfalls, and production systems characteristics. The analysis distinguishes between the short-term impacts, which are limited to the irrigation year, and long-term impacts, which are irreversible.

Based on the Mhiri and Bousnina study, the current irrigation practices cause an annual loss equivalent to about 1 percent of irrigated agricultural production. Eighty percent of this loss is short-term, and 20 percent is permanent. In an irrigated area of about 375,000 hectares,[6] this corresponds to an equivalent annual loss of 3,000 hectares and an irreversible loss of 750 hectares. Considering that the value added of irrigated agriculture is US$2,160 per hectare (Ministry of Agriculture 2005a, 2005b), the short-term losses are estimated at about US$6.5 million, and the present value of permanent losses at US$25.3 million. Consequently, the total impact of salinity and waterlogging on irrigated agriculture is estimated at US$31.8 million.

Impact of insufficiently treated wastewater on irrigated areas. Tunisia has made impressive progress in the reuse of treated wastewater. Irrigated agriculture accounts for the primary use of treated wastewater, comprising 35 million cubic meters, or 65 percent of the total volume (MARH 2004). The *Office National de l'Assainissement* (National Office of Sanitation, or ONAS) has accumulated long experience in wastewater treatment. However, the average quality of treated wastewater did not meet the required standards in 25 out of 78 total treatment stations (ONAS 2004). This substandard treated wastewater may compromise its reuse in agriculture, thus affecting the surrounding environment, surface water, and groundwater.[7]

Irrigation with low-quality treated wastewater causes substantial productivity loss in agricultural areas.[8] In 2004, 7,440 hectares (or 19 perimeters) were irrigated with treated wastewater, of which 6,066 hectares used insufficiently treated wastewater (ONAS 2004; MARH 2005e). The perimeters of Borj Touil (3,800 hectares) and Mornag (1,000 hectares) form about 80 percent of the irrigated surface using insufficiently treated wastewater.

MARH (2005c) estimates the loss of agricultural production due to poor-quality treated wastewater in Borj Touil at about US$2.6 million, or US$680 per hectare. MARH (2005d) values the loss in Mornag at US$1.1 million, or US$1,100 per hectare. Averaging the two estimates (US$890 per hectare), the productivity loss on the other lands irrigated with treated wastewater (1,050 hectares) is estimated at US$0.9 million. Adding up the estimated damages in Borj Touil, Mornag, and other perimeters, the total value of damage is US$4.6 million.

This loss is not due solely to the poor quality of treated wastewater but mainly to other factors, such as lack of drainage and current agricultural practices. No study is available that analyzes the impact of these factors on the agricultural productivity of each irrigated perimeter. Assuming that the insufficiently treated wastewater accounts for only 30 percent of the total loss,[9] its value is estimated at about US$1.4 million.

Impact of dam sedimentation. MARH (2003) projected that the storage capacity of 26 operating dams in 2002 would gradually diminish because of increased sedimentation—from 450 million cubic meters in 2002 to 800 million cubic meters in 2020, corresponding to an annual loss of 19.4 million cubic meters of dam storage capacity (MARH 2003). Dam sedimentation can cause additional costs to replace the lost storage capacity and reduce agricultural production because less water is available for irrigation (photos 2.1 and 2.2).

Replacement cost—Different procedures can be adopted to prevent or reduce reservoir sedimentation, such as antierosive treatment, heightening dams, dredging, and building dams for replacement. The unit cost of replacement varies between US$0.004 per cubic meter (for heightening) and US$0.99 per cubic meter (for building a new dam), as presented in table 2.2. Using these estimates, the total cost of replacing the dam capacity annually lost to sedimentation is valued at US$13 million.

Reduced agricultural production—If the dam capacity lost to sedimentation is not replaced, that can reduce water availability for other uses. Because agriculture is the most important water use in Tunisia, this section estimates the impact of sedimentation on irrigated agriculture. In 2004, 409 million cubic meters of dam capacity were allocated for the intensive agriculture of 85,500 hectares (Ministry of Agriculture 2005b), corresponding to about 4,800 cubic meters per hectare. Assuming that reduced water availability results in a smaller irrigated area, the annual dam sedimentation (19.4 million cubic meters) would cause a loss of 4,000 hectares of irrigated area. The loss on 1 hectare is assumed equal to the difference between the added value of irrigated and nonirrigated agriculture, estimated at US$1,520 (Ministry of Agriculture 2005a, 2005b). Therefore, the annual damage cost from the lost 4,000 hectares is US$6 million.

In the absence of investments to replace the lost dam capacity, sedimentation in 2004 would reduce water availability for irrigation during dry years over the next 25 years. Assuming cycles of dry years similar to

Photo 2.1   Reservoir in Northwest Tunisia

Source: G. Sterk.

Photo 2.2   Gullies Drain into a Reservoir in Northwest Tunisia

Source: G. Sterk.

those during 1980–2004,[10] the present value of the agricultural loss due to sedimentation in 2004 is about US$29 million.

Because dam sedimentation does not affect agricultural productivity during rainy years, and certain dams cannot be replaced, the actual cost of dam sedimentation varies, between US$13 million and US$29 million. This is just a conservative estimate of the dam sedimentation impacts.[11]

Table 2.2   Cost of Replacing Dam Capacity

Replacement procedures	Capacity (m³/year millions)	Cost (US$/m³)	Total cost (US$/year millions)
Dam heightening[a] (x)	6.34	0.004	0.03
Building new dams[b] (y)	13.10	0.99	12.99
Dredging[c]	not estimated	very high	not estimated
Antierosive treatment[d]	not estimated	not estimated	24
Total replacement cost (x + y)	19.44	n.a.	13

Sources: DGETH 2003; MARH 2003.

Notes: m³ = cubic meters, n.a. = not applicable.

a. The heightening of the Sidi Salem dam increased reservoir capacity by 264 million cubic meters, at a total cost of US$1 million, corresponding to an average of US$0.004 per cubic meter. According to this estimate, the cost to replace the lost capacity due to annual sedimentation (6.34 million cubic meters) is about US$25,000.

b. The cost of building new dams varies from US$0.45 per cubic meter (for the Harka dam) to US$1.45 per cubic meter (for the Moula dam), averaging US$0.99 per cubic meter (General Directorate for Large Hydraulic Works, pers. communication, 2006).

c. This procedure is extremely expensive and poses problems in selecting the discharge zone.

d. To reduce dam sedimentation, the *Direction Générale de l'Aménagement et de Conservation des Terres Agricoles* (General Directorate of Planning and Conservation of Agricultural Land, or DGACTA) planned a project including soil and conservation works to protect the watershed, with a total budget of US$240 million over the 2002–11 period, which is equivalent to US$24 million per year.

Overall costs. The total impacts over 25 years of the water pollution and sedimentation that occurred in 2004 on irrigated agriculture is conservatively estimated to cost US$46 million to US$62 million. The impact of salinity and waterlogging on irrigated agriculture is the most important impact, accounting for nearly 60 percent of the average estimated cost.

Fishing

National fish production increased from 83,600 tons to 109,600 tons, or by 31 percent, during the 1995–2004 period because of fleet development (CGP 2005). The evolution at the national level, however, masks a decreasing trend at the local level.

In the Gabès and Sfax governorates, fish production diminished from 35,300 tons in 2001 to 31,100 tons in 2005, corresponding to a reduction in the governorates' joint share of national production from 36 percent to 28 percent (MARH 2005a). Although these estimates cover a short period of time, several publications have indicated that the Sfax and Gabès governorates are the areas most affected by water pollution (SCET 2002; SOGREAH 2002; INSTM 2006). Therefore, this analysis estimates the impacts of water degradation in those two areas only. Consequently, the analysis underestimates the nationwide impacts of pollution on fishing by not capturing the possible effects of water pollution in other governorates.

No data indicating the cause-and-effect relationship between water degradation and fishing were found in Tunisia. However, the impact of water degradation on fish production can be approximated by comparing the fish production trend in the polluted areas (Sfax and Gabès) with production in the rest of the country. Figure 2.1 indicates that fish production in the rest of the country increased steadily over the decade of 1995–2004, whereas production in the polluted areas reached a ceiling and declined during the latter half of the decade, 2001–04.

Figure 2.2 shows the trend in fish production per boat during the same period. Throughout Tunisia, excluding the polluted areas, production increased from 7.5 tons to 12 tons per boat, but it remained at 8 tons per boat in the polluted areas. A simple regression indicates that if the fish production had followed the same trend in the polluted areas as in the rest of the country, the production in the polluted areas would have been 13.6 tons per boat instead of 8 tons per boat.

Considering that 3,900 fishing boats used the polluted areas in 2004 (MARH 2005c) and that the value of fish was US$2,480 per ton (MARH 2005b), the forgone benefit is US$54.2 million. This loss results from several factors in addition to water pollution, such as fishing practices and lack of investments. Lacking more accurate information, about 20–30 percent

Figure 2.1   Fish Production in Most Polluted Areas vs. Rest of Tunisia

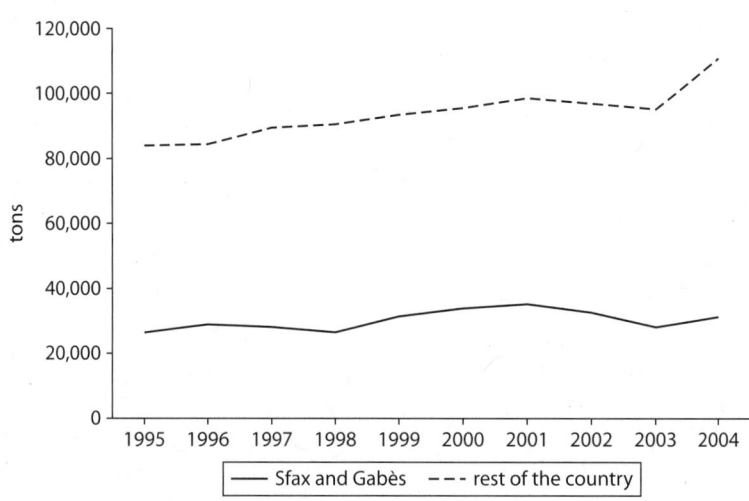

Source: Authors, based on data from CGP 2005.
Note: Fish production in the "rest of the country" is Tunisia's national fish production minus the production in the Sfax and Gabès governorates.

Figure 2.2   Fish Production per Boat in Most Polluted Areas vs. Rest of Tunisia

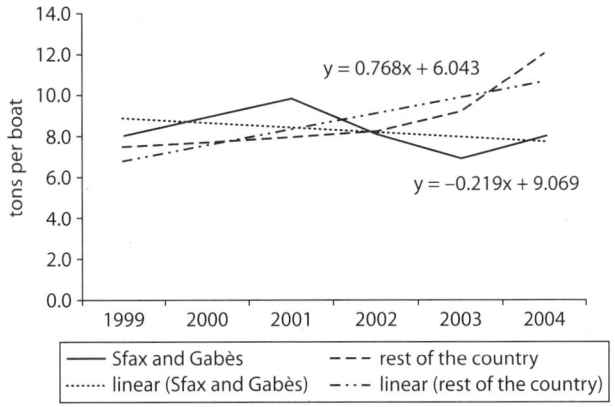

Source: Authors, based on data from CGP 2005 and MARH 2005a.
Note: Fish production in the "rest of the country" is Tunisia's national fish production minus the production in the Sfax and Gabès governorates.

of this loss is attributed to pollution.[12] Thus, the annual estimated cost of water pollution is between US$11 million and US$16 million.

Health

Access to drinking water supply and sanitation services in Tunisia is relatively high compared with other Middle Eastern and North African countries. Access to drinking water is 88 percent in rural areas and 100 percent in urban areas (SONEDE 2005). Access to sanitation services is well developed in urban areas, where 82 percent of wastewater is collected, of which 92 percent is treated (ONAS 2005).

In contrast, rural areas are not yet properly endowed with sanitation services. Inadequate water supply, sanitation, and hygiene affect health through waterborne diseases such as diarrhea, hepatitis A, and typhoid fever. For each of these diseases, we estimate the social costs of mortality and morbidity, the direct costs of treating morbidity cases, and the opportunity costs of time spent by caregivers. Because the estimates for hepatitis A and typhoid fever are negligible, only the valuations of the impacts of inadequate drinking water supply, sanitation, and hygiene on diarrhea cases are presented below.

Under-five mortality. Once an important cause of death for children under five years old, diarrhea has been successfully controlled in the past decade. The under-five mortality rate diminished from 52 per thousand to 25 per thousand live births during 1990–2004 (UNICEF 2004).[13] Child mortality due to diarrhea declined from as much as 7 percent of child deaths in 2000 (WHO 2006) to as little as 1.5 percent in 2004.[14] This percentage of total under-five mortality corresponds to about 62 child deaths from diarrhea in 2004.

Not all mortality cases due to diarrhea are imputable to inadequate supply of drinking water, sanitation facilities, and hygiene. Available information in Tunisia is insufficient to establish a cause-and-effect relationship between these factors and the number of diarrhea cases. However, adequate drinking water and improved sanitation and hygiene would reduce the frequency of cases by 85 percent (Hutton and Haller 2004). Assuming that only 85 percent of diarrheal cases in Tunisia are due to inadequate drinking water, sanitation facilities, and hygiene, this corresponds to about 53 such deaths of children under five in 2004. Considering that one child death is equivalent to 33 DALYs (WHO 2006), the associated damage cost is estimated at about 1,750 DALYs.

Under-five morbidity. The *Direction des Soins de Santé de Base* (Directorate of Basic Health Care, or DSSB) reported that public health centers registered 65,855 diarrheal cases among children younger than five years old (DSSB 2004). This number represents about 28 percent of all diarrheal cases in Tunisia (UNICEF 2000).[15] Therefore, the total number of diarrheal cases among children under five is estimated at 235,200. If one assumes (a) that only 85 percent of the cases are due to inadequate drinking water supply, sanitation facilities, and hygiene (Hutton and Haller 2004); (b) an average of three days' duration per case; and (c) a weighting factor of 0.12 DALY per episode (WHO 2006), the annual loss is estimated at 200 DALYs.

Over-five mortality. No information was found concerning mortality from diarrhea among individuals over five years old. If one conservatively assumes that all subjects are treated and survive, this loss corresponds to zero.

Over-five morbidity. DSSB (2004) reports 118,300 cases registered in the public sector. This number represents only 30 percent of all diarrheal cases among individuals over five (UNICEF 2000). Based on this information, we calculate a total of about 394,000 cases of diarrhea. Assuming that 85 percent of the cases are due to inadequate drinking water supply, sanitation facilities, and hygiene (Hutton and Haller 2004), and assigning a weighting factor of 0.09 DALY per episode (WHO 2006),[16] the annual loss due to morbidity in individuals over five years old is 250 DALYs.

Overall morbidity and mortality costs. Overall, the annual losses in Tunisia due to morbidity and mortality caused by inadequate water supply, sanitation, and hygiene are estimated at 2,200 DALYs. One DALY loss is valued using the human capital approach as a lower bound and the value of statistical life (VSL) as an upper bound. Applying the human capital approach, and assuming that the value of 1 DALY corresponds to the gross domestic product (GDP) of US$2,820 per capita in 2004, the estimated damage cost is US$6.2 million. No study on the VSL was found for Tunisia. However, adjusting the 2003 Viscusi and Aldy estimate[17] for the United States to Tunisia, the VSL would be equivalent to US$460,000, or US$19,000 per DALY. Accordingly, the morbidity and mortality costs correspond to US$42 million. By applying the two methods, we obtain a damage cost range of *US$6.2 million to US$42 million.*

Treatment and opportunity cost of time. Treatment costs vary according to the gravity of the case. They are estimated at US$1.8 million for children under five and at US$800,000 for individuals over five. The opportunity cost of time spent by caregivers with small children is estimated at US$100,000. The total treatment and opportunity cost of time reaches *US$2.7 million.*

If one adds the morbidity and mortality costs, the treatment costs, and the opportunity cost of time, the country's diarrheal cases related to inadequate drinking water supply, sanitation facilities, and hygiene exact a total cost of *US$9 million to US$45 million.*

Tourism
If one accounts for 5.5 percent of GDP in 2004, tourism is a growing sector in Tunisia, especially on the coast, which receives more than 5 million tourists annually (MAERH 2003). However, urbanization, uncontrolled development of resorts, and the pollution of marine water are causing a decline of recreational value, especially on the eastern shore.

Impacts of seawater degradation on national tourism. The impacts of seawater pollution on national tourism are estimated based on national tourists' additional expenditures to travel and stay in cleaner resorts.

The beaches most affected by water pollution are Sfax, Gabès, Grand Tunis, Khniss Sayada, and Soliman (SCET 2002; SOGREAH 2002).

Although no information is available concerning tourist departure rates[18] from these zones due to seawater pollution, a 2005 survey by the *Office National du Tourisme Tunisien* (Tunisian National Tourism Office, or ONTT) reveals the following:

- Departure rates from nonpolluted areas such as Bizerte, Nabeul, and Sousse range between 40 percent and 50 percent; we assume the same range for the polluted beaches, had pollution not been there.
- Current departure rates are 62 percent for Grand Tunis, 72 percent for Sfax, 52 percent for Gabès, 70 percent for Khniss Sayada, and 70 percent for Soliman.

Based on the figures above, the departure rates attributable to seawater pollution are 12–22 percent for Grand Tunis, 22–32 percent for Sfax, 2–12 percent for Gabès, 20–30 percent for Khniss Sayada, and 20–30 percent for Bizerte.

The ONTT survey (2005) also shows the tourist distribution according to the origin and destination of travel. For example, 393,000 tourists annually leave from Sfax, of whom 4 percent go to Echaffar beach (Sfax), 17 percent to Mahdia, 19 percent to Sousse, 14 percent to Hammamet, and 7 percent to Djerba. Assuming that only 22–32 percent of departures are due to seawater pollution and that the average number of trips per tourist is two, the total cost of transport and stay in cleaner resorts is estimated at US$11.7 million to US$17 million.[19] Based on the same approach, the overall cost of transport and stay due to water pollution in Sfax, Gabès, Grand Tunis, Khniss Sayada, and Soliman is estimated at *US$14.2 million to US$22.4 million.*[20]

Impacts of seawater degradation on international tourism. In 2004, European and American tourists spent about 23 million bed-nights on the coast, especially in Djerba-Zaezis, Sousse-Kairouan, Nabeul-Hammamet, and Monastir.[21] A survey carried out in Hammamet, Sousse, and Mahdia estimates that 4 percent of the international tourists are willing to pay US$13 per person per night for improved water quality and beach cleanliness and reduced beach congestion. Considering the same WTP for 4 percent of the European and American tourists, the total WTP to improve seawater quality is *US$12 million.*

Overall, the impacts of seawater degradation on national and international tourism are estimated at *US$26.2 million to US$34.4 million.*

Biodiversity

Tunisia has more than 250 wetlands, many of which provide rich habitats for biodiversity (Karem, Maamouri, and Mohamed 1999). For example, Ichkeul Lake is unique in North Africa, hosting about 124,400 migratory bird species during winter 2005–06 (DGF 1997). However, industrial and agricultural pollution are degrading many wetlands. Other wetlands are simply drying out as a result of drought and water overuse.

No relevant studies estimating the biodiversity loss in wetlands were found in Tunisia.[22] The costs of wetland degradation are estimated on the basis of the funds allocated annually for biodiversity restoration in the affected areas.[23] The annual budget allocated for building and maintaining locks to control water salinity in inland wetlands is about US$117,600 (ANPE 2007). The annual costs linked to rehabilitation of coastal and marine wetlands—Ariana, Mahdia, Kelibia, Slimane, and other sensitive coastal areas—are about US$5.7 million (MEAT 2002). Overall, the annual costs associated with wetland degradation are estimated to be about *US$5.8 million.* However, because of weaknesses concerning the data and methodology, this particular estimate should be regarded with extreme caution.

Overexploitation of Groundwater

Groundwater is essential for Tunisia, a country scarce in surface water. Available resources are estimated at 772 million cubic meters of shallow aquifers and 1.4 billion cubic meters of deep aquifers (DGRE 2005b). Table 2.3 shows that, at the national level, shallow aquifers appear only slightly overexploited, with an exploitation rate of 104 percent. Deep aquifers seem underexploited, at a rate of 80 percent. However, overexploitation of deep aquifers is alarming at the *regional* level, particularly in the southwest (117 percent), and overexploitation of shallow aquifers is extremely high in the northeast and especially central-east (150 percent) (DGRE 2005a, 2005b; World Bank 2007).

Table 2.3   Tunisian Groundwater Resources and Their Exploitation

Water source	Available resources (m³ millions)	Exploited resources (m³ millions)	Exploitation rate (% exploited/ available)
Shallow aquifers	772	803	104
Deep aquifers	1,411	1,127	80

Sources: DGRE 2005a, 2005b.
Note: m³ = cubic meters.

Table 2.4 shows the volume and the change in water table for shallow and deep aquifers by region. The estimation of average change in groundwater table at the regional and national levels is based on weighting the water volume in each aquifer by the associated annual change in the water table. At the national level, shallow aquifers subject to overexploitation provide a volume of 430 million cubic meters and experience an average decline in the water table of 0.38 meters per year. Similarly, overexploited deep aquifers provide a volume of 924 million cubic meters and suffer an annual decline in the water table of 0.74 meters.

The cost of overexploitation includes the additional pumping cost due to the decline in water table and the cost of building new replacement wells.[24]

Pumping cost. The additional pumping cost necessary to extract water from a deeper level is estimated in table 2.5. For pumping of 1 cubic meter of groundwater, the average fuel consumption is about 0.004 liters per meter of depth (World Bank 2005). The cost of electric pumping is estimated to be 70 percent of the cost of diesel pumping under the same conditions of depth and volume, according to interviews with experts at the *Direction Générale des Grands Travaux Hydrauliques* (General Directorate of Water Resources, or DGRE). The annual pumping cost therefore amounts to US$1.5 million. Using a discount rate of 4 percent, the total pumping cost over 25 years is *US$23 million.*

Replacement cost of wells. According to the state of wells surveyed in 2004, the total cost of 48 wells with a total depth of 14,591 meters amounts to

Table 2.4   Water Volume and Change in Groundwater Table, by Region

Region	Shallow aquifers Overexploited aquifers (m^3 millions)	Shallow aquifers Average annual change in water table (m)	Deep aquifers Overexploited aquifers (m^3 millions)	Deep aquifers Average annual change in water table (m)
Grand Tunis	22	−0.17	25	−0.17
Northeast	252	−0.36	27	−0.40
Northwest	9	−0.27	11	−0.92
Central East	23	−0.28	5	−0.57
Central West	124	−0.48	105	−0.51
Southeast	n.a.	n.a.	174	−0.80
Southwest	n.a.	n.a.	577	−0.80
Tunisia	**430**	**−0.38**	**924**	**−0.74**

Source: Authors' compilation based on data from DGRE 2004.
Note: m = meter, m^3 = cubic meters, n.a. = not applicable.

Table 2.5   Annual Costs to Pump Additional Water for Resources Subject to Overexploitation in Tunisia

	Shallow aquifer (430 million m³)		Deep aquifer (924 million m³)	
	Diesel pumping	Electric pumping	Diesel pumping	Electric pumping
Resources subject to overexploitation[a] (m³ millions)	215	215	462	462
Average diesel consumption (liter/m of depth/m³)	0.004	n.a.	0.004	n.a.
Average annual drop in water table (m)	0.38	0.38	0.74	0.74
Market price (US$/liter of diesel)	0.512	n.a.	0.512	n.a.
Annual pumping cost (US$)	167,300	117,100[b]	700,170	490,100[b]
Total annual pumping cost (US$)	1,475,000			
Net present value (US$)	23,040,000			

Sources: Authors' compilation based on data from DGR 2004 (see table 2.4) for resources subject to overexploitation and average annual drop in water table; World Bank 2005 for average diesel consumption; local market price for diesel.
Notes: In the net present value (NPV) calculation, t = 25 years, r = 4 percent. m = meter, m³ = cubic meters, n.a. = not applicable.
a. Based on General Directorate of Water Resources (DGRE), the electrification rate for pumping is about 50 percent.
b. Seventy percent of the diesel cost (DGRE interviews, pers. communication, 2007).

Table 2.6   Replacement Cost of Wells, by Region and Nationally

	Private wells			Public wells			Total cost (US$ thousands)	Cost to replace all abandoned wells (US$ thousands)
Region	Number of wells	Depth (m)	Cost of wells (US$ thousands)	Number of wells	Depth (m)	Cost of wells (US$ thousands)		
G. Tunis	45	4,405	711	4	460	149	860	645
Northeast	174	17,453	2,819	13	2,058	665	3,483	2,613
Northwest	33	3,022	488	14	2,028	655	1,143	857
C. East	33	3,475	561	8	2,192	708	1,269	952
C. West	95	12,645	2,042	17	4,302	1,390	3,432	2,574
Southeast	30	2,811	454	29	7,926	2,560	3,014	2,261
Southwest	10	1,326	214	40	9,606	3,103	3,317	2,488
Tunisia	**420**	**45,137**	**7,290**	**125**	**28,572**	**9,229**	**16,518**	**12,389**

Sources: Authors' calculations; DGRE (2005a and b) for the number and depth of public and private wells.
Note: m = meter.

US$4.7 million. This corresponds to an average value of US$323 per meter of depth. The cost of building public wells is based on that estimate. The cost of building a private well is half of that for public wells, or US$161 per meter of depth, based on data from the Ministry of Agriculture.

Table 2.6 shows the replacement costs of public and private wells at the regional and national levels. The replacement cost of private wells is US$7.3 million, and that of public wells is US$9.2 million, amounting to a total cost of US$16.5 million. According to the Ministry of Agriculture, 75 percent of these wells are built to replace wells abandoned due to the decline in water table, at an estimated cost of *US$12.4 million.*

Total cost caused by groundwater overexploitation. Based on the annual costs of pumping additional water (US$23 million) and replacing wells (US$12.4 million), the total cost due to groundwater overexploitation over 25 years is estimated at *US$35.4 million.*

Conclusions

The overall cost of water degradation and groundwater overexploitation in Tunisia is estimated at *US$165.8 million, or 0.6 percent of GDP in 2004.*[25] Figure 2.3 shows that the greatest cost of environmental damage occurs in the agricultural sector, mainly because of the impacts of salinity and waterlogging on irrigated agriculture. Groundwater overexploitation is the second-largest economic loss, because of the costs of pumping additional water and building new wells to address a

Figure 2.3   Annual Cost of Water Degradation in Tunisia

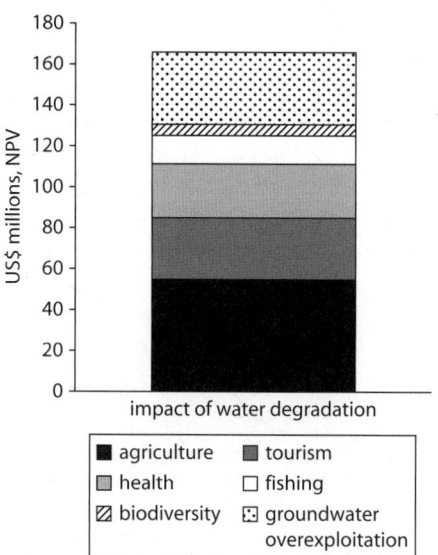

Source: Authors' calculations.
Note: NPV = net present value (t = 25 years, r = 4 percent).

decline in the water table. Other significant damages are the impacts on tourism of polluted seawater and the impacts on health of water-borne diseases caused by inadequate water quality, sanitation, and hygiene practices.

The results show that the priority interventions are (a) to improve agricultural practices in order to reduce the impact of salinity and (b) to diminish the groundwater overexploitation. The core issues are to find the most appropriate means to limit the negative effects of salinity and to devise instruments able to reduce the groundwater overexploitation. The latter is especially important in sensitive regions such as Southern Tunisia, Central Tunisia, and the Cap Bon Peninsula, where groundwater represents a critical resource. Among the available options are ways to artificially recharge groundwater with conventional water and treated wastewater, taking the necessary measures—in terms of treatment and selection of recharge areas—to avoid contamination.

In addition to investments in infrastructure, a wide variety of options exist to remediate different forms of water degradation in Tunisia, including the following:

- Strengthening control of pollution sources
- Valuing and reinforcing incentives and aid for pollution cleanup (for example, through government-administered *Fonds de Dépollution*)
- Providing tariff incentives to support water savings
- Expanding reuse of treated wastewater, especially in Grand Tunis[26]
- Improving coordination among institutions for better management of groundwater (for example, among ONAS; the National Agency for Environmental Protection; and the Ministry of Agriculture)
- Investing in water sector and other projects that will improve the efficient management of water resources in Tunisia [27]

In addition, cost-benefit analyses of the various options are important to prioritize the areas of intervention, to find the least expensive combination of investments, and to incorporate the decisions within an integrated water management policy.

Notes

Lelia Croitoru is an Environmental Economist at the World Bank, Middle East and North Africa Region, 1818 H Street, NW, Washington, DC 20433, USA.

Maria Sarraf is a Senior Environmental Economist at the World Bank, South Asia Region, MC10-1019 1818 H Street, NW Washington, DC 20433, USA.

Fadhel Ghariani is an Environmental Specialist, IDEA-Tunisie, 6 Rue du Mali - 1002 Tunis, Tunisia.

Mohamed Salah Matoussi is a Professor at the University of Tunis, Faculty of Economic Sciences and Management, El Manar 2001, Tunis, Tunisia.

Hamed Daly-Hassen is a Researcher at the Institut National de Recherches en Génie Rural, Eaux et Forêts (INRGREF) B.P. 10, 2080 Ariana, Tunisia.

1. Water in Tunisia is naturally saline. It is important to distinguish between natural water salinity and the increased salinity level from inappropriate irrigation practices. This chapter focuses only on the latter.

2. *Chott* is a dry (salt) lake in the Saharan area of Africa that stays dry in the summer but receives some water in the winter. http://en.wikipedia.org/wiki/Chott.

3. The sustainable use of water can be defined in several ways. In this chapter, the sustainability level is considered achieved when the current water use does not threaten future water availability.

4. This method gives a common value to disease and premature deaths caused by environmental degradation in terms of DALYs.

5. The duration of the selected time horizon is identified with a generation life span, under the assumption that a person of average age would enjoy the benefits of the environment for another 25 years, until death.

6. The total irrigated area is estimated based on data from Mhiri and Bousnina (1999), updated to reflect the expansion rate of irrigated lands from 1999 through 2005.

7. In fact, the irrigation of the Moknine perimeter (104 hectares) was stopped because of the bad quality of treated wastewater.

8. The treated wastewater not reused and discharged could degrade the environment, especially in coastal zones and particularly in the Gulf of Tunis. The impact on tourism of discharging treated wastewater was incorporated in the section about the impacts on tourism.

9. Based on interviews with wastewater experts during the preparation of the report on the cost of environmental degradation (World Bank 2007).

10. The dry years observed during this period include 1987–89, 1993–95, and 2000–01 (M. Louati, Ministry of Agriculture, pers. communication).

11. In addition to the estimated impacts, sedimentation causes other impacts such as eutrophication and increase of water turbidity, which have not been estimated.

12. Based on a 2007 interview with M. S. Romdhane, president of the National Commission of Planning and Evaluation of Research Results (Fish and Aquaculture).

13. UNICEF (United Nations Children's Fund) Tunisia statistics, 2004. http://www.unicef.org/french/infobycountry/Tunisia_statistics.html.

14. The lower-end estimate is based on a 2006 interview with Dr. R. Maktouf, Programme national de lutte anti-diarrhéique.

15. The rest of the diarrheal cases comprise the number of cases inventoried at the National Social Security Fund (2 percent), in the private sector (20 percent), and those who do not approach a doctor (50 percent) (UNICEF 2000). The last group includes those who use the advice of pharmacists or friends, use self-medication, or are not treated at all.

16. The World Health Organization (WHO) provides similar weighting factors for these age groups: 5–14 (0.094), 15–44 (0.086), 45–59 (0.086), and over 60 (0.088) (WHO 2006). We use the average value of 0.09.

17. Viscusi and Aldy (2003) estimated the value of a statistical life in the United States to range between US$5.5 million and US$7.6 million in 2000 prices.

18. The departure rate represents the ratio of the number of interviewed people stating that they spent at least one night out of their usual environment for purposes not linked to paid activities (such as recreation, visits to parents and friends) to the total number of interviewed people.

19. The cost of transport to other unpolluted zones assumes the cheapest alternative of four people per car and US$0.4 per kilometer of gas. Estimating the cost of stay assumes that (a) if the unpolluted destination cities are closer than 200 kilometers from the origin city, we consider that tourists stay only one night, at a low-season cost of US$12 per night per person; and (b) if the unpolluted destination cities are farther than 200 kilometers from the origin city, we consider that tourists stay an average of 2.3 nights per trip, at a low-season cost of US$40 per night per person (ONTT 2005).

20. This is a conservative estimate because (a) it does not include the WTP of residents of polluted areas who cannot travel to other areas; (b) it does not consider other polluted sites, such as the area around Oued Hamdoun; and (c) it does not capture the one-day visitors because of lack of information.

21. Tourism-related statistical information comes from the National Statistics Institute. http://www.ins.nat.tn.

22. However, a wide range of studies have been carried out worldwide on estimating the value of wetlands (Brower and others 1997; Acharya 2000; Barbier 2000; Woodward and Wui 2001; Schuyt 2005; Birol, Karousakis, and Koundouri 2006). Because of the different conditions involved, it is not meaningful to transfer the results of these studies to Tunisia.

23. Use of the replacement cost method may be problematic. When potential rather than actual expenditures are used, it is not always clear whether the environmental benefit in question justifies the costs of replacing the damage

(Bishop 1999). Some argue that, in most cases, the replacement costs tend to overestimate the value of damage (Heal 1999).

24. Because of data insufficiency, the estimation does not include the value of deepening and rehabilitating existing wells. Moreover, the cumulative impact of the annual overexploitation of groundwater may contribute to reducing water availability for future generations. The objective of this chapter is to estimate only the impacts of the overexploitation occurring in 2004. We estimate this value using the additional cost of pumping and replacement cost of wells.

25. The total cost is stated in terms of net present value (t = 25 years, r = 4 percent discount rate).

26. The Ministry of Environment and Sustainable Development is undertaking a study concerning the transfer of treated wastewater from Grand Tunis to irrigate the industrial biodiesel crop production.

27. Such projects include Tunisia's *Projet d'Investissement dans le Secteur de l'Eau* (Water Sector Investment Project, or PISEAU).

References

Acharya, G. 2000. "Approaches to valuing the hidden hydrological services of wetland ecosystems." *Ecological Economics* 35 (1): 63–74.

ANPE (National Agency for Environmental Protection). 2007. "Rapport sur le suivi scientifique au parc national de l'Ichkeul, 2005–2006." Unité chargée du suivi de l'Ichkeul, Ministry of Environment and Sustainable Development, ANPE, Tunis.

Barbier, E. B. 2000. "Valuing the Environment as Input: Applications to Mangrove-Fishery Linkages." *Ecological Economics* 35 (1): 47–61.

Birol, E., K. Karousakis, and P. Koundouri. 2006. "Using Economic Valuation Techniques to Inform Water Resources Management: A Survey and Critical Appraisal of Available Techniques and an Application." *Science of the Total Environment* 365 (1–3): 105–122.

Bishop, J. T. 1999. *Valuing Forests: A Review of Methods and Applications in Developing Countries*. London: International Institute for Environment and Development.

Brower R., I. H. Langford, I. J. Bateman, T. C. Crowards, and R. K. Turner. 1997. "A Meta-Analysis of Wetland Contingent Valuation Studies." Centre for Social and Economic Research on the Global Environment (CSERGE) Working Paper GEC 97–20, CSERGE, University of East Anglia, Norwich, U.K.

CGP (Commissariat Général de la Pêche). 2005. *Base de données internes*. Tunis: CGP.

DGETH (General Directorate for Large Hydraulic Works). 2003. "Gestion de l'envasement dans les retenues des grands barrages tunisiens." Rapport élaboré par Mme ABID Madiha, MARH, Tunis.

DGF (General Directorate of Forests). 1997. "Conservation et développement de la flore et de la faune sauvage et des aires protégées," 2ème pdf, Ministère de l'Agriculture.

DGRE (General Directorate of Water Resources). 2004. *Annuaire piézométrique de Tunisie.* Tunis: DGRE.

————. 2005a. *Annuaire d'exploitation des nappes profondes de Tunisie.* Tunis: DGRE.

————. 2005b. *Situation de l'exploitation des nappes phréatiques de Tunisie.* Tunis: DGRE.

DSSB (Directorate of Basic Health Care). 2004. *Bulletin épidémiologique.* Tunis: DSSB.

Heal, G. 1999. "Valuing Ecosystem Services." Working Paper Series in Money, Economics and Finance. Columbia Business School, Columbia University, New York.

Hutton, G., and L. Haller. 2004. *Evaluation of the Costs and Benefits of Water and Sanitation Improvements at the Global Level.* Geneva: World Health Organization.

INSTM (National Institute of Marine Sciences and Technologies). 2006. "Report of Activities 2002–2006" for the Republic of Tunisia, INSTM, Tunis.

ITES (Tunisian Institute for Strategic Studies). 2002. "The Future of Water: A New Challenge for Tunisia." ITES, Tunis.

Karem A., F. Maamouri, and A. Ben Mohamed. 1999. *Actes du séminaire "Gestion et conservation des zones humides tunisiennes."* Proceedings, Projet MedWet, Sousse, Tunis.

Lebdi, F. 2005. *Contraintes de l'agriculture irriguée aux opportunités du marché: Cas de la Tunisie.* Cairo: Regional Office for the Near East, Food and Agriculture Organization of the United Nations (FAO).

MAERH (Ministry of Agriculture, Environment, and Water Resources). 2003. *Gestion de l'envasement dans les retenues des grands barrages tunisiens.* Tunis: MAERH.

MARH (Ministry of Agriculture and Water Resources). 2003. *Gestion de l'envasement dans les retenues des grands barrages tunisiens.* Rapport élaboré par Mme ABID Madiha. Tunis: MARH.

————. 2004. *Note sur le développement de la réutilisation des eaux usées traitées dans les irrigations des périmètres irrigués et des espaces verts.* Tunis: MARH.

————. 2005a. *Annuaire statistique de l'agriculture et de la pêche.* Tunis: MARH.

————. 2005b. *Bilan économique de l'année 2005 pour l'agriculture et la pêche.* Tunis: MARH.

————. 2005c. *Etude de diagnostic et de réhabilitation du périmètre de Cebala Borj Touil irrigué à partir des eaux usées traitées.* SERAH/DHV Consultant BV. Tunis: MARH.

————. 2005d. *Etude de diagnostic et de réhabilitation du périmètre de Mornag irrigué à partir des eaux usées traitées.* SERAH/DHV Consultant BV. Tunis: MARH.

————. 2005e. *Planification et gestion de l'eau en Tunisie.* Tunis: MARH.

Mathers, C., C. Bernard, K. Iburg, M. Inoue, D. Ma Fat, K. Shibuya, C. Stein, N. Tomijima, and H. Xu. 2004. "Global Burden of Disease in 2002: Data Sources, Methods, and Results." Revised. Global Programme on Evidence for Health Policy Discussion Paper 54, World Health Organization, Geneva.

MEAT (Department of Environment and Spatial Planning). 2002. *Xème plan de développement économique et social: Environnement et aménagement du Territoire.* Tunis: MEAT.

MEDD (Ministry of Environment and Sustainable Development). 2007. *Projet d'Investissement dans le Secteur de l'Eau* (Water Sector Investment Project, or PISEAU). *Etude relative à l'actualisation de l'inventaire des principales sources potentielles de pollution des ressources hydriques et de la mise en place d'un réseau national de surveillance de la pollution hydrique.* Tunis: MEDD.

Mhiri, A., and H. Bousnina. 1999. "Diagnostic agro-environnemental de l'état des terres cultivées dans les divers systèmes de production en Tunisie." *Revue de l'INAT* (*Journal of the National Institute of Tunisian Agronomy,* or INAT), Centenaire de l'INAT, Séminaire "Ressources naturelles."

Ministry of Agriculture. 2005a. *Enquête sur les périmètres irrigués 2004.* Tunis: Ministry of Agriculture.

————. 2005b. *Planification et gestion de l'eau.* Tunis: Ministry of Agriculture.

ONAS (National Office of Sanitation). 2004. *Rapport annuel d'exploitation des stations d'épuration.* Tunis: ONAS.

————. 2005. *Rapport annuel d'exploitation des stations d'épuration 2004.* Tunis: ONAS.

ONTT (National Office of Tunisian Tourism). 2005. *Etude du marché touristique tunisien.* Rapport Phase 1, janvier 2005, ACC. Tunis: ONTT.

OTED (Tunisian Observatory for Environment and Sustainable Development). 2004. *Etat de l'environnement 2004.* Tunis: OTED.

Sarraf, M., B. Larsen, and M. Owaygen. 2004. "Cost of Environmental Degradation: The Case of Lebanon and Tunisia." Environment Department Paper 97, World Bank. Washington, DC.

SCET (Central Society for Equipment Planning). 2002. *Plan d'occupation des plages zones centre et sud—plage de Gabès*. Tunis: SCET.

Schuyt, K. 2005. "Economic Consequences of Wetland Degradation for Local Populations in Africa." *Ecological Economics* 53 (2005): 177–90.

SOGREAH Consultants. 2002. *Protection des ressources marines et côtières du Golfe de Gabès*. Tunis: SOGREAH.

SONEDE (National Water Exploitation and Distribution of Water). 2005. *Rapport statistique année 2004*. Tunis: SONEDE.

UNICEF (United Nations Children's Fund). 2000. *Multiple Indicator Cluster Survey: Tunisia*. New York: UNICEF.

Viscusi, W. K., and J. E. Aldy. 2003. "The Value of a Statistical Life: A Critical Review of Market Estimates throughout the World." *The Journal of Risk and Uncertainty* 27 (1): 5–76.

WHO (World Health Organization). 2006. "Global Burden of Disease." Health statistics and health information systems, WHO. *http://www.who.int/health info/global_burden_disease/en/index.html*.

Woodward, R., and Y-S. Wui. 2001. "The Economic Value of Wetland Services: A Meta-Analysis." *Ecological Economics* 37 (2): 257–70.

World Bank. 2004. "Tunisia: Country Environmental Analysis (1992–2003)." Report 25966-TN, Water, Environment, Rural and Social Development, Middle East and North Africa Region, World Bank, Washington, DC.

———. 2005. *Islamic Republic of Iran: Cost Assessment of Environmental Degradation*. Sector Note, Report 32043-IR, Middle East and North Africa Region. Washington, DC: World Bank.

———. 2007. "République Tunisienne: Evaluation du Coût de la Dégradation de l'Eau." Report 38856-TN, Department of Sustainable Development, Middle East and North Africa Region, World Bank, Washington, DC.

CHAPTER 3

Air Pollution: The Case of Jordan

Lelia Croitoru, Raffaello Cervigni, and Amer Jabarin

Air pollution is an environmental challenge faced by Jordan (Government of Jordan 2005). Transport, power generation, and industry account for the bulk of air emissions in the country (AFD 2006). Air pollution is a growing problem, particularly in heavily populated cities such as Amman, Zarqa, and Irbid, although not in centers of tourism such as the Dead Sea and Beida.

In the hot spots where it occurs, air pollution likely affects health significantly, through airborne diseases, and the environment, through reduced visibility and aesthetic value of landscape. This chapter estimates these impacts in the most polluted areas of Jordan. It is based on a more extensive study estimating the cost of environmental degradation (COED) carried out within the Jordan Country Environmental Analysis (World Bank, forthcoming).[1]

Overview of Air Pollution

Jordan has made significant progress in reducing air pollution. The country adopted regulations[2] and strategies to reduce emissions that have a negative impact on the environment—setting ambient air-quality standards comparable to international standards[3] and limiting values for industrial emissions.[4] The transport sector strategy addresses environmental concerns related to road freight and air transport (Government of Jordan 2008).[5] The revised national energy strategy sets a 6 percent target for use of renewable energy by 2020 and emphasizes the need for increased energy efficiency, which would reduce energy requirements by approximately 20 percent (Government of Jordan 2007). The country adopted cleaner fuels in the energy and transport sectors.[6] Moreover, in 2008, Jordan started to phase out leaded gasoline and high-sulphur diesel to improve fuel quality in order to meet the European (Euro 4) emission standards.[7]

Despite these achievements, available evidence indicates that in selected hot spots of industrial activity and vehicular traffic, air quality is poor. Table 3.1 shows that vehicles represent a significant source of emissions in Jordan. The vehicle fleet is rapidly increasing at an annual rate of 7 percent to 10 percent.[8] Nevertheless, the vehicle fleet in relatively old, with about 33 percent of the vehicles produced before 1990. Old cars are still maintained and used, contributing significantly to emissions. Because Amman and South Amman are home to about 69 percent of all Jordanian vehicles (AFD 2006), they represent major hot spots for air pollution.

Emissions from the industrial sector originate mainly from the cement plants in Fuheis and Rashidyia; the industrial area of Hashimyeh near Zarqa; power plants and the phosphate and potash industries in Aqaba; and other sources. Among these, mining contributes the most to air pollution—accounting for about 62 percent of the total suspended particulates (TSP), 78 percent of the particulate matter less than 10 microns in diameter (PM_{10}), and 39 percent of the nitrogen oxides generated by industry.[9]

Table 3.1   Pollutant Emissions from Transportation and Industry Sources, 2004

Source of emissions	NO_x	SO_2	TSP
Road transportation (%)	79[a]	20[b]	39
Other diesel (%)	0	0	1
Air transport (%)	2	1	14
Industry (%)	7	30	18
Electricity production (%)	11	48	28
Total (tons)	**72,900**	**123,000**	**6,500**

Sources: AFD 2006; World Bank, forthcoming.
Notes: Percentages represent the share of pollutants emitted by a source relative to other transportation and industry sources. NO_x = nitrogen oxides, SO_2 = sulphur dioxide, TSP = total suspended particulates.
a. Percentage calculated based on the average of emissions estimates, ranging between 53,063 tons (AFD 2006) and 62,160 tons (World Bank, forthcoming).
b. Percentage calculated based on the average of emissions estimates, ranging between 24,240 tons (AFD 2006) and 25,994 tons (World Bank, forthcoming).

Several institutions are in charge of monitoring air pollutants in Jordan, such as the Ministry of Environment in Hashimyeh and the Ministry of Health in the Greater Amman Municipality (GAM). However, the country has no centralized repository of information on ambient air quality and emissions. Most of the information made available to the study team about pollutant sources and emissions comes from the French Development Agency (AFD 2006), completed with data from the Ministry of Environment. The lack of easily available information makes it difficult to reach a comprehensive understanding of the air pollution situation in the country.

Methodology

The cost of discomfort from air pollution is based on benefit transfer of the results of a contingent valuation study carried out in Morocco. Estimating the health impacts of air pollution involves their clear identification and valuation through the several steps and methods presented below. Valuation of all the impacts discussed in this chapter refers to the year 2006.

Identifying the Health Impacts

Although PM_{10} is thought to cause substantial health damages, particulate matter smaller than 2.5 microns in diameter ($PM_{2.5}$) have more significant health effects (Pope and others 2002). Sulphur dioxide (SO_2) and nitrogen oxides (NO_x) may also have important consequences because they can react with other substances in the atmosphere to form particulates.

Estimated in this chapter are the following impacts of $PM_{2.5}$ and PM_{10} in the GAM, Zarqa, Aqaba, Fuheis, and Rashadeia and the effects of SO_2 in Zarqa:

- Infant and child mortality related to respiratory diseases caused by short-term exposure to PM_{10}
- Adult mortality related to cardiopulmonary diseases and lung cancer caused by long-term exposure to $PM_{2.5}$
- All-age mortality related to exposure to SO_2
- All-age morbidity related to exposure to PM_{10}, including chronic bronchitis, hospital admissions of patients with respiratory problems, emergency room visits, restricted activity days, lower respiratory infections in children, and general respiratory symptoms.

Valuation steps. Valuation is based on four steps, which are articulated in the following section:

1. Identifying pollutants and measuring concentration
2. Estimating the population exposed
3. Establishing the dose-response coefficients
4. Estimating the health effects.

Methods to measure the health impacts (physical valuation). The health impacts of air pollution are estimated in terms of disability-adjusted life years (DALYs). This methodology has been developed and applied by the World Health Organization (WHO) and the World Bank in collaboration with international experts to provide a common measure of disease burden for various illnesses and premature mortality.[10]

The DALY method weighs illnesses by severity: a relatively mild illness or disability represents a small fraction of a DALY, and a severe illness represents a larger fraction of a DALY. Mortality due to health problems also is expressed in terms of DALYs: a year lost to premature mortality represents 1 DALY, and the future years lost are discounted at a fixed rate of 3 percent. Morbidity is expressed in terms of DALYs and other costs of illness. Sarraf (2005) offers further detail on the use of the DALYs method.

Methods to estimate the health impacts (monetary valuation). The value of 1 DALY can be estimated using two approaches. The human capital approach (HCA) estimates the indirect cost of productivity loss through the value of an individual's future earnings (Kirch 2008). Accordingly, 1 DALY corresponds to one person's average contribution to production, namely the gross domestic product (GDP) per capita. This method provides a lower bound for the loss of 1 DALY.

The second approach, the value of a statistical life (VSL), measures the willingness to pay (WTP) to avoid death. This estimation is reached by observing individual behavior when trading off health risks and money (Johansson 2006). The VSL is calculated by dividing the marginal WTP to reduce the risk of death by the size of the risk reduction. By this measurement, the value of 1 DALY corresponds to the VSL divided by the number of discounted average years of life lost because of an adult's death (World Bank 2005). The VSL method provides an upper bound of health damages.

In addition to the costs of pain and suffering, society incurs direct costs related to illness. These costs are computed through the cost of illness approach (COI). The COI estimates the treatment costs linked to different health end points (for example, hospitalization, restricted activity days, and doctor visits) and the cost of caregivers' time to treat sick individuals (that is, the caregiver's wage).

Cost of Air Pollution

This section estimates the impacts of air pollution on health in terms of mortality and morbidity due to $PM_{2.5}$, PM_{10}, and SO_2 exposure, and on the environment, through losses of landscape in Jordan's major cities.

Cost of the Health Effects of Air Pollution

The analysis focuses on the health effects attributable to $PM_{2.5}$ and PM_{10} exposure in the GAM, Zarqa, Aqaba, Fuheis, and Rashadeia and the impact of SO_2 exposure in Zarqa. Because the GAM accounts for about half of the country's population, and the valuation for other areas is similar to that for the GAM, this section describes in detail the health effects of $PM_{2.5}$ and PM_{10} in the GAM only and presents the estimation results for all of the areas considered.

Step 1: Identify pollutants and measure concentration. Table 3.2 presents the estimated average PM_{10} concentrations in the selected areas,

Table 3.2   Estimated PM_{10} Concentration and Exposed Population in Selected Urban Areas

Location	Estimated average PM_{10} concentration ($\mu g/m^3$)	Exposed population (millions)
GAM	67[a]	1.40
Zarqa	95[b]	0.54
Fuheis	58[c]	0.01
Rashadeia	54[d]	0.01

Source: World Bank, forthcoming. Estimations of PM_{10} concentrations are based on observations reported by the Ministry of Environment, Ministry of Health, or Royal Scientific Society of Jordan. Estimations of exposed populations are based on data from the Department of Statistics and on coefficients of exposure from expert local opinion.
Note: GAM = Greater Amman Municipality. $\mu g/m^3$ = micrograms per cubic meter.
a. Ministry of Health data for Al-Husseini, annual average 2002–07.
b. Royal Scientific Society of Jordan data for the Ministry of Environment, 2003–04, based on observations taken every other day.
c. Ministry of Environment, annual average 2004–06.
d. Ministry of Environment, annual average 2003–06.

based on the main sources of information. Estimating the average PM_{10} concentration in the GAM involves the following steps:

1. The Al-Husseini station in downtown Amman is selected as a reference for what is likely to be the highest concentration of PM_{10} throughout Amman. Downtown Amman corresponds to Al-Madinah, one of the 27 districts of the GAM. The Ministry of Health has been measuring the PM_{10} concentration at Al-Husseini over the period 2002–07. Based on these measurements, the average annual concentration of PM_{10} over this period is estimated to be 124 micrograms per cubic meter.

2. The PM_{10} concentrations for the other districts are estimated by using the PM_{10} concentration in Al-Madinah and a specific scaling factor, varying between 0 and 100 percent, that reflects the potential impact on reference PM_{10} concentration (the Al-Husseini station measurement) of wind, topography, and traffic in each district, based on local expert opinion (staff of the Ministry of Environment and the GAM).

3. The average PM_{10} concentration in the GAM is calculated on the basis of the PM_{10} concentration and the population density in each of its districts. The weighted average concentration for the GAM as a whole is estimated at 67 micrograms per cubic meter.

Step 2: Estimate the population exposed. For each GAM district, the exposed population is estimated by multiplying the total resident population by a coefficient of exposure. The latter reflects the proportion of the total population that spends most of the time in the district where they reside, based on expert local opinion (staff of the Ministry of Environment and the GAM). The average total exposed population in the GAM is the sum of the exposed population in each district and accounts for 64 percent of the GAM's total population. In the absence of more information, the coefficient of exposure for the other areas is assumed to be equal to that for the GAM. Table 3.2 presents the estimated exposed population for each area considered.

Step 3: Establish dose-response coefficients. The impacts of PM_{10} and $PM_{2.5}$ on *mortality* are estimated based on the relative risk (RR) functions provided below (Ostro 2004). Because $PM_{2.5}$ data are not available for Jordan, we approximate them by converting PM_{10} levels using a factor of

0.6 (Cohen and others 2004). We use the threshold levels of 10 micrograms per cubic meter for $PM_{2.5}$ and 20 micrograms (μg) per cubic meter for PM_{10}, as provided by the WHO air quality guidelines (WHO 2005).[11]

a. For mortality related to short-term exposure of children under 5 years of age:

$$RR = \exp[\beta\,(x - x_0)], \tag{3.1}$$

where β ranges between 0.0006 and 0.0010,

x = current annual mean concentration of PM_{10} (μg per cubic meter),

x_0 = baseline concentration of PM_{10} (μg per cubic meter).

b. For cardiopulmonary mortality related to long-term exposure of adults over 30 years of age (Pope and others 2002):

$$RR = [(x + 1)/(x_0 + 1)]^\beta, \tag{3.2}$$

where β ranges between 0.0562 and 0.2541,

x = current annual mean concentration of $PM_{2.5}$ (μg per cubic meter),

x_0 = baseline concentration of $PM_{2.5}$ (μg per cubic meter).

c. For lung cancer mortality related to long-term exposure of adults over 30 years of age (Pope and others 2002):

$$RR = [(x + 1)/(x_0 + 1)]^\beta, \tag{3.3}$$

where β ranges between 0.08563 and 0.37873,

x = current annual mean concentration of $PM_{2.5}$ (μg per cubic meter),

x_0 = baseline concentration of $PM_{2.5}$ (μg per cubic meter).

To estimate the impacts of PM_{10} on *morbidity*, we consider the following health end points: chronic bronchitis, hospital admissions of patients with respiratory problems, emergency room visits, restricted activity days, lower respiratory infections in children, and general respiratory symptoms. Table 3.3 presents the dose-response coefficients for morbidity due to PM_{10} exposure.

Step 4: Estimate the health effects. The health effects of air pollution are converted to DALYs to facilitate comparisons with health effects from

Table 3.3   Dose-Response Coefficients for Morbidity from PM$_{10}$ Exposure
per 1 µg/m^3 annual average ambient concentration

Annual health effect	Dose-response coefficients
Chronic bronchitis[a]	0.9
Respiratory hospital admissions[b]	1.2
Emergency room visits[b]	23.5
Restricted activity days[a]	5,750
Lower respiratory illness in children[c]	169
Respiratory symptoms[a]	18,300

Sources: Ostro 1994; Abbey and others 1995.
Note: µg/m^3 = micrograms per cubic meter.
a. Per 100,000 adults.
b. Per 100,000 population.
c. Per 100,000 children.

other environmental factors and comparisons between mortality and morbidity using a common indicator.

Table 3.4 presents the number of DALYs per 10,000 cases of various health end points and the estimated total number of DALYs. Overall, the total cost of mortality and morbidity caused by PM$_{2.5}$ and PM$_{10}$ exposure in the GAM, Zarqa, Aqaba, Fuheis, and Rashadeia, and by SO$_2$ exposure in Zarqa totals 15,670 DALYs.

The cost of adult mortality is estimated based on the HCA as a lower bound and the VSL as an upper bound, thus obtaining a wide range. Applying the HCA, 1 DALY corresponds to the annual GDP per capita, or US$2,510, in 2006 (Central Bank of Jordan 2007). Using the VSL method, the value of 1 DALY in Jordan is equivalent to US$18,310 after adjusting for the GDP per capita differences between the United States and Jordan.[12] Accordingly, the cost of adult mortality is estimated in the range of US$13 million to US$252 million, averaging to US$133 million.[13]

The cost of child mortality and all morbidity is estimated based only on the HCA because data are scarce in Jordan concerning the WTP of individuals to avoid pain and discomfort. Accordingly, this cost is estimated at US$15 million. In addition, the COI is estimated at US$9 million.

Adding up the estimated costs of mortality, morbidity and the COI, the total cost of health impacts in Jordan associated with PM$_{10}$, PM$_{2.5}$, and SO$_2$ exposure is estimated in the range of *US$37 million to US$276 million*, averaging to *US$156.5 million*.

Table 3.4   Estimated Loss of DALYs from Health Effects of Air Pollution

Annual health effect	DALYs lost per 10,000 cases [a]	Total DALYs		
		Minimum	Maximum	Average
Mortality due to PM$_{10}$ exposure (children under 5)	80,000	100	880	490
Mortality due to PM$_{2.5}$ exposure (adults over 30)	80,000	2,860	11,280	7,070
Mortality due to SO$_2$ exposure[14]	80,000	2,480	2,480	2,480
Total mortality (a)		**5,440**	**14,640**	**10,040**
Chronic bronchitis (adults)	22,000	1,890	1,890	1,890
Respiratory-related hospital admissions	160	30	30	30
Emergency room visits	45	160	160	160
Restricted activity days	3	1,650	1,650	1,650
Lower respiratory illness (children)	65	590	590	590
Respiratory symptoms	0.75	1,310	1,310	1,310
Total morbidity (b)		**5,630**	**5,630**	**5,630**
Mortality and morbidity (a + b)		**11,070**	**20,270**	**15,670**

Sources: World Bank 2005; Larsen 2004; authors' calculations. Adapted from the World Bank, forthcoming.
a. Worldwide mortality DALYs (World Bank 2005); morbidity DALYs (Larsen 2004).

Cost of Discomfort from Air Pollution

In addition to its impacts on health, air pollution also reduces visibility and the aesthetic value of landscapes. Data on such impacts are not available in Jordan. A contingent valuation study of urban households in Rabat-Salé, Morocco, found a WTP for a 50 percent reduction in air pollution of about US$13 to US$15 per household per month (Belhaj 2003). About 10 percent of this figure is thought to express the cost of discomfort from air pollution, the rest reflecting the effects on health. Adjusting for GDP per capita differentials between Morocco and Jordan and considering about 300,000 households in Amman and Zarqa, the cost of discomfort in the urban areas of Jordan reaches *US$4 million to US$5 million*, averaging to *US$4.5 million*.

Conclusions

The total cost of damage from air pollution in Jordan averages about *US$161 million, or 1.15 percent of GDP in 2006*. The health effects—especially from adult mortality tied to pollutant exposure—account, by far, for the largest component of this cost, as figure 3.1 illustrates:

- Adult mortality due to $PM_{2.5}$ exposure is the single most important damage contributor, accounting for 66 percent of the total cost.
- Adult mortality due to $PM_{2.5}$ and SO_2 exposure account for more than 80 percent of the total cost.
- Morbidity due to PM_{10} exposure and the cost of illness together account for just 15 percent of the total cost.

It should be noted that several estimates suffer from data constraints. In some cases, the absence of local information required the use of benefit

Figure 3.1   Estimated Damages from Air Pollution

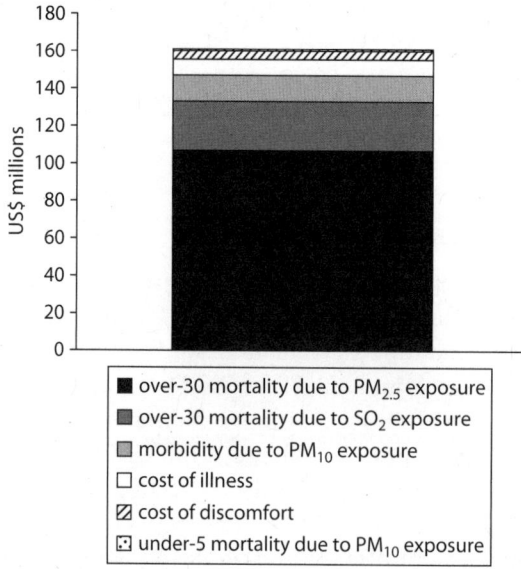

Source: Adapted from World Bank (forthcoming).

transfer from international studies, such as when valuing the impact of air pollution on the visibility and aesthetic value of landscapes. In other cases, the data limitations were so severe that some losses could not be estimated at all. The major issues not covered by this chapter include the impact of pollutants other than $PM_{2.5}$, PM_{10}, and SO_2 on health; the impact of $PM_{2.5}$ and PM_{10} on health in polluted areas other than the GAM, Zarqa, Aqaba, Fuheis, and Rashadeia; and the impacts of SO_2 in polluted areas other than Zarqa. For these reasons, the estimated cost of air pollution should be regarded as an order of magnitude.

The World Bank (forthcoming) estimates the total COED in Jordan at US\$334 million, or 2.35 percent of GDP. As figure 3.2 reflects, air pollution (which is highly localized) accounts for almost half of the COED. The relatively high cost of air pollution should be interpreted as a signal of degradation in selected road-traffic and industrial hot spots (such as downtown Amman, Zarqa, and Fuheis) and *not* as a reflection of overall poor air quality at the national level. In particular, *air pollution does not affect Jordan's tourism poles* such as Petra, the Dead Sea, and Jerash. Moreover, the magnitude of the problem in Jordan is less significant than in comparator countries in the region (for example, the Arab Republic of Egypt) or elsewhere (for example, Peru).

Based on an extensive analysis of the main sources of air pollution in Jordan (the transport and industrial sectors), the World Bank (forthcoming)

Figure 3.2   Shares of Environmental Degradation Cost in Jordan, by Category

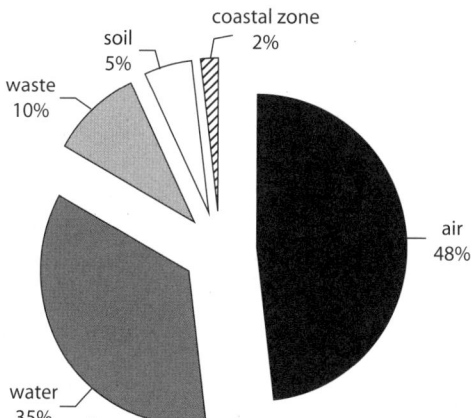

Source: Adapted from World Bank (forthcoming).
Note: Percentages represent the estimated share of US\$334 million (total annual cost of environmental degradation in Jordan) attributable to each damage category.

suggests several actions to reduce air pollution in the hot spots. Short-run recommendations include the following:

- Accelerate enhancement of the air quality monitoring system.
- Speed up the phase-out of high-sulphur diesel fuel.
- Define an action plan for implementing other key policies to reduce emission per unit of traffic.[15]
- Establish Jordan-specific emission factors to improve information about the volume and distribution across the vehicular fleet of polluting emissions from road transport.[16]

Over a longer period, the proposed recommendations include (a) the establishment of traffic models to optimize transportation system design and the related benefits of reducing air pollution and congestion, and (b) the definition of air pollution abatement plans.[17]

Notes

Lelia Croitoru is an Environmental Economist at the World Bank, Middle East and North Africa Region, 1818 H Street, NW, Washington, DC 20433, USA.

Raffaello Cervigni is a Lead Environmental Economist and Climate Change Regional Coordinator at the World Bank, Africa Region, 1818 H Street, NW, Washington, DC 20433, USA.

Amer Jabarin is an Associate Professor, University of Jordan, Department of Agricultural Economics and Agribusiness, Amman, Jordan.

1. Chapter 3 of the Country Environmental Analysis updates the 2005 study prepared for the Mediterranean Environmental Technical Assistance Program, assessing the cost of environmental degradation of each natural resource, (METAP 2005).

2. Environmental Protection Law No. 1, 2003.

3. The ambient air quality standards (JS 1140, updated in 2006) provide limits for total suspended particulates (TSP) and PM_{10} (particulate matter with a diameter smaller than 10 microns) and for gaseous substances sulphur dioxide (SO_2), carbon oxide (CO), nitrogen oxides (NO_x), hydrogen sulphide (H_2S), and lead (Pb).

4. The standards for emissions from stationary sources (JS 1189, updated in 2006) sets limits for TSP by type of industry as well as for gaseous substances.

5. The transport strategy includes provisions to reduce impacts from road freight transport and to rationalize energy consumption; it also aims to reduce noise

and emissions from aircraft by adopting relevant international standards and upgrading national legislation.

6. Use of natural gas to replace diesel and heavy fuel oil by the power sector increased substantially, reaching 77 percent of total fuel use in 2006.

7. The Euro 4 standard for passenger cars specifies per-kilometer emissions for diesel of 0.5 grams of carbon oxide (CO), 0.3 grams of hydrocarbon (HC) and NO_x, 0.25 grams of NO_x, and 0.025 grams of PM; and for gasoline of 1 gram of CO, 0.1 grams of HC, and 0.08 grams of NO_x (http://en.wikipedia.org/wiki/European_emission_standards).

8. The percentage is based on the information provided by the license and vehicle inspection directorate and the Ministry of Environment.

9. The estimates are based on Industrial Pollution Projection System (IPPS) methodology (World Bank, forthcoming).

10. Information on the WHO Web site further defines disability-adjusted life years. http://www.who.int/healthinfo/global_burden_disease/metrics_daly/en/.

11. The threshold levels represent the baseline concentrations below which there are no health impacts.

12. The valuation uses a VSL of US$3.5 million, based on an earlier study that estimated the VSL to avoid a statistical cancer case in the United States to be within the range of US$3.2 million to US$3.7 million (Gayer, Hamilton, and Viscusi 2000). More recently, a meta-analysis of more than 60 studies valuing mortality risk premiums and 40 studies of injury risk premiums estimated the VSL within the range of US$4 million to US$9 million in United States, using labor market data (Viscusi and Aldy 2003). Thus, the VSL estimate applied in this chapter can be considered conservative.

13. The estimates are attributable to $PM_{2.5}$ and PM_{10} exposure and result from multiplying the number of DALYs (lower and upper bounds) by the estimated value of a DALY (lower and upper bounds).

14. To estimate mortality caused by SO_2, we use the WHO methodology based on the following parameters: (a) total population size (540,000 in Zarqa; Department of Statistics 2006, http://www.dos.gov.jo); (b) the average mortality rate (0.007; Ministry of Health, Annual Statistical Book 2006); (c) the population ratio under 18 years old (0.37; U.S. Census Bureau 2008, http://www.census.gov and (d) the average SO_2 concentration in the city (171 micrograms per cubic meter; based on monthly measurements in Electrical Training Center, Ibn El-Anbari and Um-Shuraik School, as provided by the Ministry of Environment). It results in a range of 129 to 780 death cases due to exposure to SO_2 every year, averaging to 309 cases. Considering 80,000 DALYs per 10,000 cases, this corresponds to about 2,480 DALYs.

15. Such policies include better maintenance of older vehicles and improved environmental performance of the vehicular fleet through strengthened regulation of imports.

16. Such emission factors would be based on driving circle-measurement or other internationally accepted methodologies.

17. See World Bank (forthcoming) for more details on these recommendations.

References

Abbey, D. E., M. D. Lebowitz, P. K. Mills, F. F. Petersen, W. L. Beeson, and R. J. Burchette. 1995. "Long-Term Ambient Concentrations of Particulates and Oxidants and Development of Chronic Disease in a Cohort of Non-Smoking California Residents." *Inhalation Toxicology* 1995 (7): 19–34.

AFD (French Development Agency). 2006. "Air Pollution Monitoring in Jordan." Air Report for the Quality Monitoring Program, Ministry of Environment, September 2006, AFD, Paris.

Belhaj, M. 2003. "Estimating the Benefits of Clean Air: Contingent Valuation and Hedonic Price Methods." *International Journal of Global Environment Issues* 3 (1): 30–46.

Central Bank of Jordan. 2007. *Monthly Statistical Bulletin.* Various issues. Periodical of the Central Bank of Jordan, Amman.

Cohen, A. J., H. R. Anderson, B. Ostro, K. Pandley, M. Krzyzanowski, N. Kunzli, K. Gutscmidt, C. A. Pope III, I. Romieu, J. M. Samet, and K. Smith. 2004. "Urban Air Pollution." In *Comparative quantification of health risks: Global and regional burden of disease due to selected major risk factors, Vol. 2,* ed. M. Ezzati M, A. D. Rodgers, A. D. Lopez, and C. J. L. Murray, 2282. Geneva: World Health Organization.

Department of Statistics. 2006.

Gayer, T., J. T. Hamilton, and W. K. Viscusi. 2000. "Private Values of Risk Tradeoffs at Superfund Sites: Housing Market Evidence on Learning about Risk." *Review of Economics and Statistics* 82 (3): 439–451.

Government of Jordan. 2005. *National Agenda.* Hashemite Kingdom of Jordan. Amman.

———. 2007. *National Energy Sector Strategy Update 2007–2020.* Amman.

———. 2008. *Draft National Transportation Strategy: Recommendations.* Amman.

Johansson, P-O. 2006. "On the Definition and Estimation of the Value of a Statistical Life." Working paper 2006-23 for the Milan European Economy Workshops, October.

Kirch, W., ed. 2008. "Human Capital Approach." In *Encyclopedia of Public Health*, 698. New York: Springer.

Larsen, B. 2004. "Cost of Environmental Damage in Colombia: A Socio-Economic Study and Environmental Health Risk Assessment." Background paper for Country Environmental Analysis, World Bank, Washington, DC.

METAP (Mediterranean Environmental Technical Assistance Program). 2005. *Cost of Environmental Degradation in the Hashemite Kingdom of Jordan*. Report for METAP, World Bank, Washington, DC.

Ministry of Health.

Ostro, B. 1994. "Estimating the Health Effects of Air Pollution: A Method with an Application in Jakarta." Policy Research Working Paper 1301, World Bank, Washington, DC.

————. 2004. *Outdoor Air Pollution: Assessing the Environmental Burden of Disease at National and Local Levels*. Environmental Burden of Disease Series, No. 5. Geneva: World Health Organization.

Pope, C. A., R. T. Burnett, M. J. Thun, E. E. Calle, D. Krewski, K. Ito, and G. Thurston. 2002. "Lung Cancer, Cardiopulmonary Mortality, and Long-Term Exposure to Fine Particulate Air Pollution." *Journal of the American Medical Association* 287 (9): 1132–41.

Sarraf, M. 2005. "Cost of Environmental Degradation: Middle East and North Africa Region." Presentation at the Environmental Economics for Development Policy course, Paris, July 19.

U.S. Census Bureau. 2008.

Viscusi, W. K., and J. E. Aldy. 2003. "The Value of a Statistical Life: A Critical Review of Market Estimates throughout the World." *Journal of Risk and Uncertainty* 27 (1): 5–76.

World Bank. 2005. *Islamic Republic of Iran: Cost Assessment of Environmental Degradation*. Sector Note, Report 32043-IR, Middle East and North Africa Region. Washington, DC: World Bank.

————. Forthcoming. "Hashemite Kingdom of Jordan: Country Environmental Analysis." Report 47829-JO. Middle East and North Africa Region, World Bank, Washington, DC.

WHO (World Health Organization). 2005. *Air Quality Guidelines: Global Update 2005*. Geneva: WHO.

CHAPTER 4

Deforestation and Forest Degradation: The Case of the Islamic Republic of Iran

Lelia Croitoru

With 8,200 plant species, of which almost 2,500 are endemic, the Islamic Republic of Iran is considered a unique country in terms of its plant diversity and genetic reserves (NBSAP 2000). Its forest diversity provides a wide range of benefits, including water and soil conservation, biodiversity, and landscape value (photos 4.1a and b).

Despite these benefits, the country's forests have been severely degraded during the past half-century. Forest clearing for agriculture, firewood, and charcoal production reduced the forest area drastically, and overgrazing and overhunting are believed to be responsible for decreasing forest quality. This chapter estimates the annual costs of deforestation and forest degradation in the Islamic Republic of Iran. It is based on a more extensive study that estimated the total cost of environmental degradation (COED) in the Islamic Republic of Iran (World Bank 2005).

Overview of the Forest Sector

With about 12.4 million hectares of forests, covering 7.4 percent of the country's area, the Islamic Republic of Iran is a low-forested country in the

Photo 4.1   Forested Areas in Rasht, Gilan Province

a. Forested watershed    b. Forest village

Source: M. Sarraf.

Middle East (SCI 2007). The per capita forest cover (about 0.2 hectares) is close to the estimated average for the Middle East and North Africa region and is only one-third of the world's average (FAO 2007). Moreover, the country's commercially usable forest area per capita is only about 0.02 hectares.[1] Five distinct forest regions can be identified as follows:

- Caspian (Hyrcanian) forests, in the country's north, cover about 1.85 million hectares, comprising the humid commercial and industrial forests, and are the only forests suitable for industrial wood production in the country.
- Arassbaran forests, in the northwest, cover about 144,000 hectares, comprising semihumid forests used mainly for fuelwood.
- Zagros forests, in the west, cover about 5 million hectares, comprising the semiarid forests important for protecting water supplies, regulating climate, and providing nonwood forest products.
- Irano-Touranian forests, in the central plateau, cover about 3.3 million hectares, comprising the arid forests important for water and soil conservation.
- Khalijo-Omanian vegetation, in the south, covers about 2.1 million hectares, comprising arid tropical forests.

Since 1962, forests have been entirely owned by the state. The Forest, Range and Watershed Organization (FRWO) at the Ministry of Jihad and Agriculture (MOJA) is responsible for forest management. Forest management plans exist only for the Caspian forests. The area currently under management is about 1.3 million hectares, and the rate of forest use is about 1.46 cubic meters per hectare (Sagheb-Talebi, Sajedi, and Yazdian 2004).

In 2002, the forests' contribution to the national economy (considering only wood products) was estimated at 1.6 percent of gross domestic product (GDP) (SCI 2004), compared with agriculture's contribution of 15 percent (World Bank 2004). Such a difference is not surprising, however, considering that the gap is similar in other Middle Eastern and North African countries.[2]

However, wood represents only a small proportion of forests' total value. A study valuing forest benefits in 18 Mediterranean countries found that wood accounts for less than 15 percent of the total forest value in most eastern Mediterranean countries of the Middle East and about 35 percent of the total forest value in the northern Mediterranean (Croitoru and Merlo 2005; Croitoru 2007a, 2008).

Forest policy in the Islamic Republic of Iran aims at forest conservation through management, afforestation, and community forestry. During the past decades, the state has given considerable attention to implementing afforestation, silviculture, road construction, and wood transport projects.

However, forest policy has largely failed to conserve the country's forests. Forest clearing for agriculture, firewood, and charcoal contributed to an estimated reduction of forest area from 19.5 million hectares to 12.4 million hectares during the past 57 years, based on FRWO data. In addition, overgrazing and overhunting are often blamed for the decreasing quality of some forests, such as in the Zagros and Kuhrud-Kohbanan mountains (WWF 2007a, 2007b). Overall, the reduction of forest size and quality are important threats to environmental sustainability and the welfare of the communities living in or near forests.

Deforestation and Forest Degradation

The literature uses the terms "deforestation" and "forest degradation" in varying ways. The term "deforestation" has been used to describe three conditions: the complete loss of forest cover; the reduction of tree crown cover below a given proportion of land cover; and the loss of primary forest alone (Pagiola 2000). "Deforestation" often describes the permanent (long-term) loss of forests, but sometimes it includes temporary loss as well.

The term "forest degradation" has also been defined in several ways, generally referring to reductions in forests' productive capacity (Watson and others 2000). In this chapter, "deforestation" refers to a complete and long-term loss of forest cover. "Forest degradation" is defined as "changes within the forest class that negatively affect the stand or site and, in particular, lower the production capacity" (FAO 1995). Accordingly, estimates of deforestation do not reflect degradation.

Deforestation

Deforestation in the Islamic Republic of Iran is caused mainly by forest conversion to agricultural and other land uses as well as by clear-cutting for timber and firewood (Islamic Republic of Iran 2001). Although it is commonly agreed that deforestation has been extensive, there is no consensus on the extent of annual deforestation in the country. Different sources report different figures on the issue, as the following examples indicate:

- Amirnejad and others (2006) report an annually deforested area of some 200,000 hectares and quote a deforestation rate of 2.3 percent in the Caspian forests and 1.1 percent in the other forest types.[3]
- FRWO indicates that total forest area diminished from 19.5 million hectares to 12.4 million hectares between 1944 and 2000, corresponding to an annual loss of about 125,000 hectares, or 0.8 percent.[4]
- The Food and Agriculture Organization of the United Nations (FAO) reports no annual forest change between 1990 and 2000, which is explained qualitatively by an increase in plantation areas combined with a low deforestation rate of natural forests (FAO 2007).

The analysis in this chapter assumes conservatively that the annually deforested area is about 125,000 hectares. This estimate should be regarded with some caution, however, because the annual deforestation rate can vary considerably over a long period. As such, the average rate calculated over a 57-year period does not necessarily reflect the deforestation rate in the most recent years.

Other deforestation figures have been reported in relation to the various forest regions. For instance, Caspian forests have shrunk from 3.4 million hectares to 1.9 million hectares over the past 22 years, an average of 45,000 hectares per year (CEP 2001b). In addition, the average annual deforestation rate in Zagros forests appears to be twice that of Irano-Touranian forests (NBSAP 2000).[5] This is equivalent to about 54,000 hectares in Zagros forests and 27,000 hectares in Irano-Touranian forests. Table 4.1 summarizes the main characteristics of the five forest regions in the Islamic Republic of Iran.

Table 4.1   Characteristics of Forests in the Islamic Republic of Iran, by Region

Forest region	Forest inventory data				Deforestation		
	Forest area (ha/ thousands)	Growing stock (m³/ha)	Annual increment (m³/ha)	Biomass (t/ha)	Deforested area (ha)	% of region's forest area	% of country's forest area
Caspian	1,900	280	2–8	100–110[b]	45,000[c]	2.4	0.3
Arassbaran	100	40–55	0.7	—	..	..	..
Zagros	5,000	14	0.4	8[b]	54,000[d]	1.1	0.4
Irano-Touranian	3,300	—	—	5[b]	27,000[d]	0.8	0.2
Khalijo-Omanian	2,100	—	—	2[b]	..	..	..
Total	**12,400**	**48[a]**	**n.a.**	**60[a]**	**125,000[e]**	**1.0**	**1.0**

Sources: Sagheb-Talebi, Sajedi, and Yazdian 2004 for forest area, growing stock, and annual increment figures. Other sources as noted below.
Notes: ha = hectares, m³/ha = cubic meters per hectare, — = not available, t/ha = tons per hectare, .. = negligible.
a. FAO 2007.
b. NBSAP 2000.
c. CEP 2001b.
d. Authors' calculations, based on a 1-to-2 ratio between the deforested area of Irano-Touranian and Zagros forests, drawn from NBSAP 2000.
e. Estimate based on FRWO data.

Forest Degradation

In addition to deforestation, forests in the Islamic Republic of Iran are largely affected by degradation. This is primarily due to past and present wood overexploitation, overgrazing, and sometimes overhunting. These threats are found in the entire country but with different intensities in each forest zone. The Caspian forests alone can be divided into three broad groups based on their state of degradation (Islamic Republic of Iran 2001):

- *Very degraded forests:* These forests cannot regenerate naturally, and the stands cover less than 100 cubic meters per hectare (about 496,000 hectares).
- *Degraded forests:* These forests are covered by young masses of saplings, and the stands cover 100 to 200 cubic meters per hectare. Operations to replant these forests are needed (about 480,000 hectares).
- *Good and high-quality forests:* The stands exceed 200 cubic meters per hectare (about 856,000 hectares).

In the Zagros forests, firewood overexploitation and overgrazing have been major degrading factors. Due to the scarcity of pasture lands, more than 14.6 million livestock graze in forest areas (about 2.9 heads per hectare of forest), endangering wildlife due to food shortages, soil degradation, and erosion (Sagheb-Talebi, Sajedi, and Yazdian 2004).

Methodology

Valuation is based on the total economic value (TEV) framework, which divides forest benefits into these categories:

- *Direct use values* such as timber, firewood, and nonwood forest products (NWFPs)
- *Indirect use values* such as water protection and purification, carbon sequestration, and nutrient absorption by soil
- *Non-use values* such as the option value of pharmaceuticals and biodiversity conservation.

Valuation is based on the standard techniques shown in table 4.2. The choice of valuation methods depends largely on data availability. For benefits for which sufficient data are available, demand curve approaches are used, ranging from direct methods based on market pricing to indirect methods such as substitute goods and production function. For example, the loss of carbon in trees is estimated based on the quantity of carbon emitted and the average price in carbon markets. The loss in plant nutrients is estimated based on the quantity of nutrients lost to deforestation and the average price of their most suitable substitute (fertilizer).

In some cases, when the data are insufficient to apply demand curve approaches, cost-based methods are used. For example, the value of lost

Table 4.2   Methods to Estimate the Value of Forest Benefits

Value type	Benefit type	Valuation method[a]
Direct use	Timber, firewood	Market price
	Recreation	Cost of travel
	Grazing, hunting	Benefit transfer[b]
	Recreation	Cost of travel
Indirect use	Loss of plant nutrients	Substitute goods
	Dam sedimentation	Production function
	Carbon loss	Market price
Non-use	Option value	Benefit transfer[b]
	Existence value	Benefit transfer[c]

Source: Authors.
a. Valuation method details in Dixon and others 1994; Pearce and Moran 1994; Garrod and Willis 2000; Merlo and Croitoru 2005.
b. Benefit transfer of results from Turkey.
c. Benefit transfer of results from a specific site in the Islamic Republic of Iran.

recreation is estimated based on the cost of travel to visit forests, had deforestation not occurred. In other cases, the data are so scarce that even cost-based methods cannot be used; whenever meaningful, we use benefit transfer of results from other studies undertaken in similar contexts. In other cases still, relevant information is not available at all, thus valuation of particular benefits cannot be provided.

Costs of Deforestation and Forest Degradation

The chapter estimates the gross losses of deforestation and forest degradation without considering the possible benefits of the alternative land uses that replace forests. It estimates the present value of the present and future annual flows of damages due to deforestation and forest degradation that occurred during 2002 and refers these values to that baseline year. It uses a discount rate of 4 percent over a time horizon of 25 years. The selected time horizon is identified with a generation life span, under the assumption that a person of average age would enjoy the benefits of the environment for another 25 years, until death.[6] The analysis uses secondary data, mostly from local sources. In the absence of secondary data, whenever meaningful, available information from international sources is used.

The analysis involves several uncertainties but suffers from information too limited to address them. Two major issues are (a) the extent of the forest benefit losses and (b) the time for forests to recover, if they do.

The extent of losses can be *complete* or *partial*, depending on how the deforestation is performed. For example, the slash-and-burn conversion of forest land to agricultural use is likely to cause a complete loss of most (if not all) forest benefits. However, if deforestation means cutting trees for timber, it may cause only a partial loss of some benefits, such as extractive uses relying on forest soil and the remaining vegetation cover (for example, fodder for grazing and other plants) and forest services (for example, watershed protection). Because of the limited knowledge concerning the magnitude of losses caused by deforestation, this chapter attempts to incorporate these valuation issues by using realistic assumptions, to the extent possible given the severe data limitations.

Costs of Deforestation: Direct Use Values
This section estimates losses of direct use values grouped in five categories: timber, firewood, grazing and other tangible NWFPs, hunting, and recreation.

Timber. Timber is provided mainly by the Caspian forests. These forests have already suffered greatly from deforestation, reducing the potential mean annual increment (MAI) from 7 cubic meters per hectare to only 3 cubic meters per hectare at present (Islamic Republic of Iran 2001). The difference of 4 cubic meters per hectare is the cumulative result of past deforestation, not of the deforestation during a single year. Therefore, we assume that any hectare of deforested land today forgoes a potential sustainable annual harvest equal to the present MAI: 3 cubic meters per hectare.

If one assumes an equal division between timber and firewood and an average stumpage price of US$150 per cubic meter of timber, the annual loss of timber on a deforested area of about 45,000 hectares of Caspian forest land is estimated at about *US$10.1 million.*[7]

Firewood. Firewood overexploitation affects primarily Caspian forests but also Zagros and Irano-Touranian forests. We assume that deforestation today forgoes future firewood benefits until the forest regenerates. In Caspian forests, the potential annual growth is about 1.5 cubic meters per hectare, and the average stumpage price is about US$30 per cubic meter. This leads to an annual future loss of firewood of about US$2 million on 45,000 hectares of forests.

In the other forests, the potential sustainable harvest is assumed to be the MAI: about 0.4 cubic meters per hectare for Zagros and about 0.7 cubic meters per hectare for Irano-Touranian forests (Sagheb-Talebi, Sajedi, and Yazdian 2004). Based on a stumpage price of about US$10 per cubic meter, the annual future loss of firewood is about US$216,000 in Zagros forests and about US$189,000 in Irano-Touranian forests.

If one aggregates the estimates of each forest region, the total annual loss of firewood is about *US$2.4 million.*

Grazing and other tangible nonwood forest products. Deforestation causes losses of vegetation cover, which prevents grazing in forests and collection of tangible NWFPs such as medicinal plants, nuts, and gum. No monetary estimates of these values were found for the Islamic Republic of Iran. However, the available literature includes a wide variety of estimates for other countries.[8] Croitoru (2007b) valued the NWFPs benefits at US$32 per hectare, based on estimates of fodder for grazing, mushrooms, honey, and other minor NWFPs in 18 countries of the Mediterranean region. The average value for the eastern Mediterranean was estimated at US$19 per hectare by weighting the national-level benefits with the forest area in each of the economies analyzed: Cyprus, Israel, Lebanon, the Syrian Arab Republic, Turkey, and the West Bank and Gaza. The average benefit for Turkey—the Islamic Republic of Iran's neighboring country—is similar to the eastern Mediterranean average (Türker, Pak, and Öztürk 2005).[9]

In the absence of estimates for the Islamic Republic of Iran, this chapter uses Turkey's estimate, considered to be the most applicable to the Islamic Republic of Iran. Of this value, about US$14 per hectare is the grazing benefit, and US$5 per hectare is the value of other minor NWFPs, such as medicinal plants and forest fruits. As previously underscored, deforestation in many areas is likely to cause a reduction, but not a complete loss, of fodder for grazing and other plants growing in forests.

If one assumes that deforestation reduces roughly by half the *per hectare* benefit of grazing and other tangible NWFPs, the corresponding loss amounts to about *US$1.2 million.*

Hunting. As in the case of grazing, no economic valuation was found concerning hunting benefits for the Islamic Republic of Iran. Therefore, we use the hunting benefit, estimated in Turkey, of about US$1 per hectare of forests (Türker, Pak, and Öztürk 2005). The original estimation was based on the permit price and license fees paid annually by hunters. Although hunting is limited to designated forest areas, the estimate is an average over the national forest area, which explains its low value. Accordingly, the annual loss of hunting benefits in the total deforested area is about *US$125,000.*

Recreation. Only one study estimating the value of recreation and ecotourism was found in the Islamic Republic of Iran (Sharifi 2004). Based on the cost-of-travel approach, the study estimates the annual recreational benefits in several forest parks to be about US$6 per visitor. The annual number of visitors to forest areas is not known. However, about 3.3 million people per year visit the Islamic Republic of Iran's coastal areas of Mazandaran and Golestan (ICM 2000). It should be noted that forest parks and other protected areas (such as Golestan National Park and Jahan Nama Protected Zone in Gorgan) are among the main tourist attractions in these areas, covering some 762,800 hectares (CEP 2001a). In the absence of more precise information, we assume that roughly half of these visitors, 1.7 million, visit these areas annually. Based on the recreational value and number of visitors, the benefit of forest parks and other protected areas is about US$13 per hectare.

Deforestation is more likely to occur in forest areas with less recreational value than in the forest parks. If one considers that the deforested areas' recreational value is only half that of the forest parks (that is, US$6.50 per hectare of forests other than parks and protected areas), the annual loss of recreational benefits on deforested areas reaches *US$812,500*.

If one aggregates the estimated costs—timber, firewood, grazing and other NWFPs, hunting, and recreation—the overall estimated losses of the forests' direct use values amount to about *US$14.7 million.*

Costs of Deforestation: Indirect Use Values

This section estimates the damages caused by deforestation to three types of indirect use values: losses of plant nutrients, dam sedimentation in downstream reservoirs, and carbon losses.

Loss of plant nutrients. Soil erosion affects about 43 percent of the country's area (FAO 1996). The soil loss is likely to contribute to a decreasing quantity of plant nutrients, which may affect production in the long run.

This valuation is based on two steps: (a) estimating the annual soil loss due to deforestation, wood overcollection, and overgrazing in forests; and (b) valuing the quantity of nutrients lost annually in the soil loss. The World Bank (1995) estimated the annual soil loss at about 1 billion tons, and FAO (1996) estimated it at about 1.5 billion tons. This study conservatively considers the lower value of the two.

Various sources argue that deforestation is a major cause of annual soil loss (for example, FAO 1996; NBSAP 2001).[10] In particular, FAO (1996) cites the overexploitation of wood as the most important cause of soil erosion in the Islamic Republic of Iran—estimating that it is responsible for about 46 percent of total erosion, the highest share among all Central Asian countries. This degree of erosion would correspond to an annual soil loss of 460 million tons because of past deforestation of 7.1 million hectares. Assuming that new deforestation would increase erosion in proportion to the area affected, the annual deforestation on 125,000 hectares would increase the soil loss by 1.8 percent. Consequently, the additional annual erosion in the country would be about 18 million tons.

No comprehensive study was found regarding the average nutrient content of soil in the Islamic Republic of Iran. However, if we assume that the soil nutrient content is similar to that in Turkey (about 0.1 percent nitrogen, 0.15 percent phosphorous pentoxide, and 0.154 percent potassium oxide), the loss of nutrients totals about 72,720 tons (Environment Foundation of Turkey 1995, quoted in Bann and Clemens 2001).

Monetary valuation is based on the substitute cost method, using as a proxy the cost of fertilizers that would be necessary to substitute for the loss of nutrients. Based on an average market price of fertilizers of about US$0.1 per kilogram, the annual cost of substituting nutrients is about *US$7.3 million.*

Dam sedimentation. The Islamic Republic of Iran has 151 dams in operation with a total reservoir capacity of 25 billion cubic meters. More than 90 percent of this water (23 billion cubic meters) is currently used for irrigation (World Bank 2004). Water from storage and diversion dams is used to irrigate 22 percent of the country's total irrigated areas (that is, 1.6 million of 7.4 million hectares). The watershed areas of dams under operation total 10.6 million hectares.

Over the past four decades, intensive deforestation, rangeland overgrazing, and other factors have resulted in soil erosion in watershed areas. Soil erosion is estimated at 30 tons per hectare per year and sedimentation at 10 tons per hectare per year (MOJA 2004). Based on the same source, dam sedimentation results in a loss of reservoir storage capacity of 236 million cubic meters per year. This is equivalent to an annual loss of 1 percent of the potential dam capacity.

The damage cost resulting from dam sedimentation is estimated in terms of the potential loss in irrigated crops. Because a significant portion of irrigated land (31 percent) in the Islamic Republic of Iran is cultivated with wheat (World Bank 2004), we focus on the potential wheat yield loss. Based on an incremental yield in irrigated areas compared to rain-fed areas of 2.33 tons per hectare (Islamic Republic of Iran 2001), a quantity of irrigation water of 4,140 cubic meters per hectare of wheat field (Soil and Water Research Institute 1997), and an average wheat price of US$170 per ton, the agricultural productivity is estimated at US$0.1 per cubic meter of irrigation water.

Assuming that dams are used at full capacity, the average annual loss of 236 million cubic meters of reservoir storage capacity results in an annual loss of around US$23.6 million. Because deforestation is only one factor contributing to dam sedimentation, we assume that it accounts for only half of this value; therefore, the associated cost of dam sedimentation is estimated at *US$11.8 million.*

Carbon loss. Deforestation results in losses of carbon stock due to wood overextraction or forest conversion to other land uses. Based on the Islamic Republic of Iran's national communication to the United Nations

Convention on Climate Change (2003), the annual net emissions of greenhouse gases from the forestry sector include the following:[11]

- About 19.5 million of tons of carbon dioxide emissions from forests and other changes in stocks of woody biomass—corresponding to the difference between 20.7 million tons of carbon dioxide *release* from wood exploitation (industrial wood harvested from commercial forests in the Caspian region and fuelwood consumption from other forests) and 1.2 million tons of carbon dioxide *uptake* due to the increase in forest area by afforestation, tree plantations around villages, and establishment of parks and green areas
- About 11.9 million tons of carbon dioxide emissions from conversion of forests to other land uses—corresponding to about 9.3 million tons of carbon dioxide immediately released from burning and about 2.6 million tons of carbon dioxide released through decay and decomposition
- About 161,560 tons of carbon dioxide–equivalent emissions of other (not carbon dioxide) gases from forest conversion through burning.

Overall, net emissions are estimated at 31.5 million tons of carbon dioxide-equivalent emissions, or about 8.5 million tons of carbon.[12] Monetary valuation is based on the shadow price method, considering the average prices currently adopted by the carbon markets. A World Bank review of the state and trends of carbon markets (2003) found that prices for certified emissions reductions in developing countries average between US$12.9 and US$18.1 per ton of carbon, depending on whether the buyer or the seller took the risk of the Kyoto protocol not being ratified.[13] Based on this price range, the annual cost of carbon losses varies from US$110 million to US$154 million, averaging to *US$132 million.*[14] This is not a loss to the country but a global damage to the international community.

Consequently, the annual loss of indirect uses affecting the Islamic Republic of Iran—loss of plant nutrients from soil erosion and loss of irrigated crops from dam sedimentation—adds up to about *US$19.1 million.* When the value of the carbon loss is added, the total loss of indirect uses reaches *US$151.1 million.*

Costs of Deforestation: Non-Use Values
This section estimates the option value of pharmaceuticals[15] and the existence value of forests in terms of biodiversity conservation.[16]

Option value of pharmaceutical products. Deforestation is likely to reduce the number of plant species with rare drug potential. Although no estimate of this value was found for the Islamic Republic of Iran, Kumar (2004) reviewed several studies valuing the plant diversity for pharmaceutical uses. The estimates vary from a range as narrow as US$0.2 to US$20.6 per hectare per year (Simpson, Sedjo, and Reid 1996) to a range as wide as US$0 to US$9,177 per hectare per year (Rausser and Small 2000).

Such estimates are scarce in most Mediterranean countries. The only estimate, of about US$6.3 per hectare of forests, was found for Turkey (Bann and Clemens 2001). Valued on the basis of the rent capture approach, this figure is quite similar to that for Mexico (Adger and others 1995).[17] Applying this estimate to the total deforested area in the Islamic Republic of Iran gives an annual loss of *US$787,500.*

Existence value. Deforestation also causes losses of existence value in the deforested area. Mayan (2000) estimates the annual benefits of biodiversity conservation at US$16 per hectare in Caspian forests, US$8 per hectare in Zagros forests, and US$16 per hectare in Irano-Touranian forests.[18] If these figures are realistic, the annual biodiversity value lost due to deforestation is about *US$1.6 million.*

When the estimated option and existence values are considered, deforestation in the Islamic Republic of Iran causes an annual loss of about *US$2.4 million* in overall non-use value.

Total Cost of Deforestation

To summarize, the annual deforestation-related losses in the Islamic Republic of Iran are estimated to total about US$168.1 million. This corresponds to an average loss of US$1,345 per hectare of deforested area. The total annual losses comprise annual flows of damage costs (US$36.1 million) and the annual stock value of carbon (US$132 million). Using a social discount rate of 4 percent, the present value of the annual flows over a 25-year time span totals *US$564.1 million.*

Adding the value of carbon, the total present value of the deforestation cost is about *US$696.1 million.* This is a conservative estimate; it does not account for certain indirect use values (such as water purification and damage due to floods), option values, and existence values (such as the cultural value of forests).

Costs of Forest Degradation

Currently, only 1.3 million hectares of forest in the Caspian Sea region are commercially exploited (Sagheb-Talebi, Sajedi, and Yazdian 2004). In these areas, the potential annual growth is estimated at 7 cubic meters per hectare, while the actual maximum annual growth is about 3 cubic meters per hectare. The difference of 4 cubic meters per hectare is mainly a result of past wood overexploitation. Considering a degradation time span of 50 years, the average annual loss is about 0.08 cubic meters per hectare. Assuming that the current overharvest will reduce the future MAI at the same rate as past overharvests, the annual loss on 1.3 million hectares is about 104,000 cubic meters.

Based on an equal distribution between timber and firewood and an average stumpage price of US$150 per cubic meter of timber and US$30 per cubic meter of firewood, this is equivalent to a total annual loss of about US$9.4 million. Using a social discount rate of 4 percent, the present value of the damage cost resulting from a loss of 0.08 cubic meters per hectare of Caspian forest over a 25-year time span totals *US$146.9 million.*

Conclusions

The total damage costs due to deforestation and forest degradation in the Islamic Republic of Iran add up to *US$843 million or 0.7 percent of GDP in 2002.* Deforestation accounts for 83 percent of this value, and forest degradation for 17 percent.

The findings suggest, as figure 4.1 illustrates, that the losses of forest services in the deforested areas are more significant than the losses of timber and other extractive values. Although the losses of forest services have been only partially valued,[19] their estimated value represents most (57 percent) of the total damage.

The overall losses due to deforestation and forest degradation affect various segments of society differently: firewood and grazing are usually losses to forest users; dam sedimentation creates costs for downstream populations; recreation losses affect visitors; and biodiversity and carbon losses affect the international community as global public goods. This distributional issue has important policy implications in the country.

The total COED in the Islamic Republic of Iran is *US$8.4 billion* or *7.4 percent of GDP* in 2002 (World Bank 2005). As figure 4.2 depicts, damages due to deforestation and forest degradation account for about 10 percent of the total COED—considerably higher than the cost of coastal zone degradation or waste collection and disposal in the country.

Figure 4.1   Annual Costs of Deforestation and Forest Degradation in the Islamic Republic of Iran, by Category

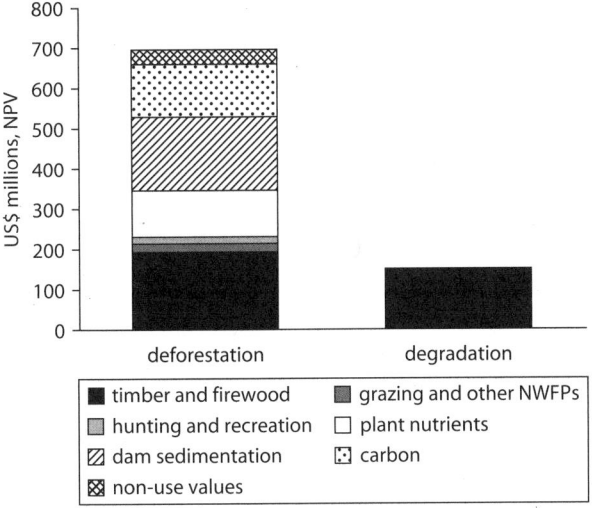

Source: Authors' calculations.
Note: NPV = net present value (t = 25 years, r = 4 percent).

Figure 4.2   Shares of Environmental Degradation Cost in the Islamic Republic of Iran, by Category

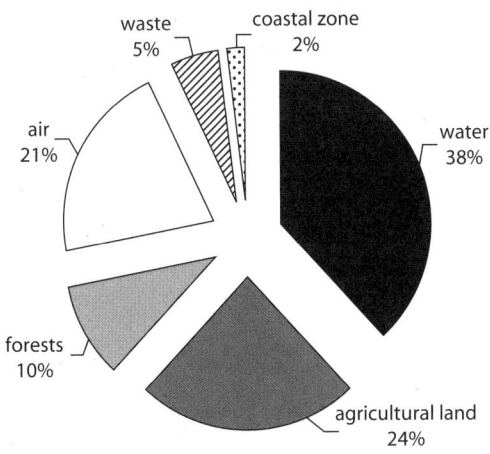

Source: Authors' calculations.
Note: Percentages represent the estimated share of US$8.4 billion (total annual COED in the Islamic Republic of Iran) attributable to each damage category.

In contrast, the percentages of the COED associated with agricultural land, air, and water are greater than the forest-related costs, mainly because of the high level of cropland salinity, the impacts of air pollution in major cities, and the health effects of water pollution.

Notes

Lelia Croitoru is an Environmental Economist at the World Bank, Middle East and North Africa Region, 1818 H Street, NW, Washington, DC 20433, USA.

1. Calculated with reference to a total population of 68.3 million inhabitants in 2006 (World Bank 2007).

2. For example, in Turkey, the forestry contribution is only 0.5 percent of GDP, compared to agriculture's 13 percent contribution (Türker, Pak, and Öztürk 2005).

3. The reported figures of Amirnejad and others (2006) are based on information from Abbasi and Mohammadzadeh (2001) and Agheli (2003).

4. Several local experts argue against the reliability of the 1944 inventory. Some of them argue that the current annual deforestation rate is about 50,000 hectares per year, or 0.4 percent of total forest area (Sagheb-Talebi, pers. communication, 2004).

5. The NBSAP (2000) reports that (1) the Zagros forest area fell from 12 million hectares to 5.5 million hectares over the past 50 years, which would correspond to an average deforestation rate of about 140,000 hectares per year; and (2) in the Irano-Touranian forest region, the area covered by juniper forests (*Juniperus polycarpus*) diminished from 3.4 million hectares to 500,000 hectares during the past 50 years, or an average deforestation rate of about 60,000 hectares per year of this forest type. These figures are assumed to be too high to be considered for the most recent years. Therefore, this paper considers the ratio of the average deforested areas between the two types of forests.

6. In reality, the losses due to deforestation may occur and persist during time periods other than 25 years. These losses may be either *permanent* or *temporary*, depending on whether forest benefits recover in time. For example, loss of timber for future uses is temporary if the forest regenerates in time, and it is permanent if an alternative use of land substitutes for the forests. In addition, the time period during which the benefits recover varies from one benefit to another, depending on several factors, such as type of benefit, climate, and the way in which the deforestation was performed.

7. The timber and firewood price estimates in this chapter are drawn from the expert opinion of M. Seifollahian, member of High Council of Forests and Forestry Organization of the Islamic Republic of Iran.

8. Lampietti and Dixon (1995) reviewed a range of studies and found values from as low as US$5 per hectare per year to as high as US$422 per hectare per year. Godoy and others (2000) found estimates of about US$18 to US$24 per hectare per year. However, these studies focused on Central American countries and none of them on Middle Eastern countries.

9. The estimated value of grazing is based on the quantity of fodder grazed in forests and the average market price of fodder's best substitute: barley (Türker, Pak, and Öztürk 2005). Collection of NWFPs is estimated based on the quantity of NWFPs collected in the forests and their market price. The estimate per hectare of forests is obtained by dividing the aggregated value of each benefit by the total forest area.

10. Overgrazing in high stocking areas has resulted in serious erosion and loss of vegetation and soil in many parts of the Caspian area, which are now sensitive to desertification (CEP 2002). Soil erosion in the Zagros mountains commonly reaches 10 tons per hectare per year (GEF and UNDP 2001). Forest conversion to agriculture and overgrazing in forests are the main causes of soil erosion in this region.

11. Aside from the figures included in this section, the Islamic Republic of Iran's national communication reports under the heading "GHG emissions from forestry sector" include two additional estimates: (a) about 0.46 million tons of carbon dioxide as net emissions from the conversion of pastures, farmlands, and other managed lands (other than forests); and (b) 0.47 million tons of carbon dioxide as net uptake from the spontaneous regeneration of grasslands and as a result of the abandonment of managed lands such as farmlands, orchards, pastures, and other lands (other than forests). Because these two additional estimates do not refer to forests, they are not considered in this chapter.

12. The calculation is based on a conversion factor of 3.66.

13. When this analysis was conducted, the Kyoto Protocol had not yet been ratified and there was some risk that it would not be, in which case carbon credits would have been worth much less.

14. In relative terms, this valuation corresponds to about 252 tons of carbon dioxide equivalent or US$1,056 per hectare of deforested area. These estimates are extremely high if compared to the per hectare losses of other forest benefits; however, they are not surprising when compared with those in other parts of the world. For example, in Latin America, clearing dense tropical forests for pasture is common practice, releasing about 500 tons of carbon dioxide per hectare (Chomitz and others 2007) and costing the society about US$2,120 per hectare, based on similar prices.

15. Option value is the value that people place on maintaining the option to benefit in the future from the forests' drug potential.

16. Existence value refers to the benefit obtained simply from knowing that certain wilderness areas or species are being conserved.

17. The approach used the following parameters: (a) the number of forest species yielding medicinal products, (b) the royalty rates that would be payable to the host country, (c) a coefficient of rent capture, (d) the likely value of internationally traded pharmaceutical products, and (e) the forest area.

18. A more recent study conducted a contingent valuation survey that estimated the average willingness to pay to protect the existence value of Caspian forests at about US$2.50 per household per month, which is equivalent to US$6.34 per person per year (Amirnejad and others 2006). However, this valuation was confined to households in important regional cities, which are not representative of the population as a whole.

19. The valuation included carbon loss, dam sedimentation, non-use values, hunting, and recreation, but it did not capture the losses of water purification and flood damages due to deforestation and the losses of forest services due to forest degradation.

References

Abbasi, A., and S. Mohammadzadeh. 2001. "Investigation of world experiences for local participation on forest resources management and utilization of successful experiences in Iran." [Farsi] In *National proceedings of North Forest Management and Sustainable Development*, vol. 2: 411–32. Tehran: Rangelands and Forests Organization.

Adger, W. N., K. Brown, R. Cervigni, and D. Moran. 1995. "Total Economic Value of Forests in Mexico." *Royal Swedish Academy of Sciences* 24 (5): 286–96.

Agheli, K. L. 2003. "Green GNP Calculation and Sustainable Degree of National Income in Iran." [Farsi with English abstract] PhD thesis, Tarbiat Modares University, Tehran.

Amirnejad, H., S. Khalilian, M. Assareh, and M. Ahmadian. 2006. "Estimating the Existence Value of North Forests of Iran by Using a Contingent Valuation Method." *Ecological Economics* 58 (4): 665–75.

Bann, C., and M. Clemens. 2001. "Turkey Forest Sector Review." Global Environmental Overlays Program Final Report, World Bank, Washington, DC.

CEP (Caspian Environment Programme). 2001a. *Habitat and Biodiversity of the Caspian Sea*. Tehran: CEP.

———. 2001b. "State and Challenges of the Marine and Coastal Environment of the Caspian Sea (Transboundary Diagnostic Analysis)." Final report (phase 2) of the Tacis CEP, CEP and EU/TACIS, Tehran.

————. 2002. *National Caspian Action Plan (NCAP) for the Islamic Republic of Iran*. Tehran: CEP.

Chomitz, K., P. Buys, G. De Luca, T. S. Thomas, and S. Wertz-Kanounnikoff. 2007. *At Loggerheads? Agricultural Expansion, Poverty Reduction, and Environment in the Tropical Forests*. Washington, DC: World Bank.

Croitoru, L. 2007a. "How Much are Mediterranean Forests Worth?" *Forest Policy and Economics* 9 (5): 536–45.

————. 2007b. "Valuing the Non-Timber Forest Products in the Mediterranean Region." *Ecological Economics* 63 (4): 768–75.

————. 2008. "Value of Mediterranean Forests." In *Encyclopedia of Earth*, ed. M. McGinley (topic) and Cutler J. Cleveland (volume). Washington, DC: Environmental Information Coalition, National Council for Science and the Environment. http://www.eoearth.org/article/Value_of_Mediterranean_forests.

Croitoru, L., and M. Merlo. 2005. "Mediterranean Forest Values." In *Valuing Mediterranean Forests: Towards Total Economic Value*, ed. M. Merlo and L. Croitoru, 37–68. Wallingford Oxfordshire, U.K.: CABI Publishing.

Dixon, J. A., L. F. Scura, L. A. Carpenter, and P. B. Sherman. 1994. *Economic Analysis of Environmental Impacts*. London: Earthscan.

Garrod, G., and K. Willis. 2000. *Economic Valuation of the Environment: Methods and Case Studies*. Northampton, MA: Edward Elgar.

GEF and UNDP (Global Environment Facility and United Nations Development Programme). 2001. "Conservation of Biodiversity in the Central Zagros Landscape Conservation Zone." Project brief for the Government of the Islamic Republic of Iran, GEF, Washington, DC; UNDP, New York. http://cbcz.ir/en/PDF/Zagros_Project_Document_En.pdf.

Godoy, R., D. Wilkie, H. Overman, A. Cubas, G. Cubas, J. Demmer, K. McSweeney, and N. Brokaw. 2000. "Valuation of Consumption and Sale of Forest Goods from a Central American Rain Forest." *Nature* 406 (6791): 62–63.

FAO (Food and Agriculture Organization of the United Nations). 1995. "Forest Resources Assessment 1990: Global Synthesis." FAO Forestry Paper 124, FAO, Rome.

————. 1996. "Population Change—Natural Resources—Environment Linkages in Central and South Asia." Report prepared for the Population and Development Service, Gender and Population Division, FAO, Rome.

————. 2007. *State of the World's Forests 2007*. Rome: FAO.

ICM (Integrated Coastal Management). 2000. "Country Profile: Iran." Research report, ICM, Andhra Pradesh, India.

Islamic Republic of Iran. 2001. *National Coastal Profile 2001*. Tehran: Islamic Republic of Iran.

————. 2003. *Initial National Communication to United Nations Framework Convention on Climate Change (UNFCCC)*. Tehran: Department of Environment, Islamic Repulic of Iran, with cooperation of the United Nations Development Programme. http://unfccc.int/resource/docs/natc/irancl.pdf.

Kumar, P. 2004. "Valuation of Medicinal Plants for Pharmaceutical Uses." *Current Science* 86 (7): 930–37.

Lampietti, J. A., and J. A. Dixon. 1995. "To See the Forest for the Trees: a Guide to Non-Timber Forest Benefits." Environment Department Paper 13, World Bank, Washington, DC.

Mayan, H. 2000. "Ecologic Valuation of Caspian Forests, Deserts and Mountains." Thesis, Department of Environment, Tarbiat Modares University, Tehran.

Merlo, M., and L. Croitoru, ed. 2005. *Valuing Mediterranean Forests: Towards Total Economic Value*. Wallingford Oxfordshire, U.K.: CABI Publishing.

MOJA (Ministry of Jihad and Agriculture). 2004. *Watershed Management*. Tehran: Forest, Rangelands and Watershed Management Organization, Islamic Republic of Iran.

NBSAP (National Biodiversity Strategy and Action Plan). 2000. "The First National Report for the Convention on Biological Diversity." Report of the NBSAP Secretariat, Tehran.

————. 2001. "Thematic Report of the Islamic Republic of Iran on Forest Ecosystems to the Convention on Biological Diversity." Report of the NBSAP Secretariat, Tehran.

Pagiola, S. 2000. "Land Use Change in Indonesia." Background paper prepared for the Environment Department, World Bank, Washington, DC.

Pearce, D., and D. Moran. 1994. *The Economic Value of Biodiversity*. London: Earthscan.

Rausser, C. G., and A. A. Small. 2000. "Valuing Research Leads: Bioprospecting and the Conservation of Genetic Resources." *Journal of Political Economy* 108 (1): 173–206.

Sagheb-Talebi, K., T. Sajedi, and F. Yazdian. 2004. *Forests in Iran*. Technical Publication No. 339-2003, Forest Research Division. Tehran: Research Institute of Forests and Rangelands.

SCI (Statistical Centre of Iran). 2004. *Yearbook: Agriculture, Forestry and Fishery*. Tehran: SCI.

————. 2007. *Iran Statistical Yearbook 1384 (2005–2006)*. Tehran: SCI. http://eamar.sci.org.ir/.

Sharifi, M. 2004. "Introduction to Eco-Tourism Development in Iran and other South and Central Asian Countries." Paper presented at Second SACAM (South and Central Asia "Man and the Biosphere") Network Meeting,

"Sustainable Eco-Tourism in Biosphere Reserves and Similarly Managed Areas," Zibakenar, Iran, September 25–28, 2004.

Simpson R. D., R. A. Sedjo, and J. W. Reid. 1996. "Valuing Biodiversity for Use in Pharmaceutical Research." *Journal of Political Economy* 104 (1): 163–85.

Soil and Water Research Institute. 1997. *Estimation of Water Needs for Main Field Crops and Horticulture, Vol. 1: Field Crops.* Tehran: Soil and Water Research Institute.

Türker, M., M. Pak, and A. Öztürk. 2005. "Turkey." In *Valuing Mediterranean Forests: Towards Total Economic Value*, ed. M. Merlo and L. Croitoru. Wallingford, Oxfordshire, U.K.: CABI Publishing.

Watson, R. T., D. Verardo, D. Dokken, I. R. Noble, and B. Bolin, eds. 2000. *Land Use, Land-Use Change, and Forestry: A Special Report of the Intergovernmental Panel on Climate Change.* Cambridge, U.K., and New York: Cambridge University Press.

World Bank. 1995. "Islamic Republic of Iran Environment Strategy Study." Report 12806-IRN, World Bank, Washington, DC.

———. 2003. *State and Trends of the Carbon Market 2003.* Washington, DC: World Bank.

———. 2004. "Islamic Republic of Iran: An Agricultural Policy Note." Report 29428-IR. Water, Environment, Social and Rural Development Department, Middle East and North Africa Region, World Bank, Washington, DC.

———. 2005. *Islamic Republic of Iran: Cost Assessment of Environmental Degradation.* Sector Note, Report 32043-IR, Middle East and North Africa Region. Washington, DC: World Bank.

———. 2007. *World Development Indicators.* Washington, DC: World Bank.

WWF (World Wildlife Fund). 2007a. "Kuhrud-Kohbanan Mountains Forest Steppe." In *Encyclopedia of Earth*, ed. Mark McGinley (topic) and Cutler J. Cleveland (volume). Washington, DC: Environmental Information Coalition, National Council for Science and the Environment. http://www.eoearth.org/article/Kuhrud-Kohbanan_Mountains_forest_steppe.

———. 2007b. "Zagros Mountains Forest Steppe." In *Encyclopedia of Earth*, eds. Mark McGinley (topic) and Cutler J. Cleveland (volume). Washington, DC: Environmental Information Coalition, National Council for Science and the Environment. http://www.eoearth.org/article/Zagros_Mountains_Forest_Steppe.

CHAPTER 5

Land Degradation: The Case of Morocco

Maria Sarraf and Abdeljaouad Jorio

In Morocco, 93 percent of the land is arid, and 7 percent is subhumid and humid (FAO 2004).[1] Soils are fragile because of their organic-matter content of less than 2 percent, even in humid zones (World Bank 2003). These soils suffer from water and wind erosion. Under these conditions, overexploitation and unsustainable management of the already vulnerable soils lead to land degradation. This chapter estimates the costs of annual land degradation in terms of the lost productivity of croplands and rangelands. It is based on an earlier, more extensive study that estimated the overall cost of environmental degradation (COED) in Morocco (World Bank 2003).

Overview of Land Resources

Agriculture is a significant sector of Morocco's economy, contributing 18 percent of gross domestic product (GDP) and employing about 33 percent of the country's active population. Both croplands and rangelands are important for Moroccan agriculture (photos 5.1 and 5.2). Arable land and permanent crops represent 9.3 million hectares, or 21 percent of the country's area. Irrigated agriculture covers about 1.4 million hectares and consumes 82 percent of the available water. Although this area is only a small proportion of the total cultivated area (15 percent), it plays

Photo 5.1   Wheat Field in Morocco

Source: S. Pagiola.

Photo 5.2   Grazing Sheep in Morocco

Source: S. Pagiola.

a disproportionately important role in agricultural production because of its high productivity and lower vulnerability to drought (FAO 2005).[2]

Of the 1.5 million holdings in Morocco, more than 70 percent have less than 5 hectares (FAO 2004). Small-farm production is usually diversified, including cereals, vegetables, and livestock. These farms have low income levels and are highly vulnerable to market forces and droughts. Morocco is among the oil-poor countries in the Middle East and North Africa, relying on cereal imports and having fiscal deficits; this makes the country vulnerable to food price shocks, which could affect the nation's inflation rate, budget, and poverty level (World Bank, FAO, and IFAD 2009).[3]

Rangelands occupy about 65 million hectares, according to the Ministry of Agriculture, Rural Development, Waters and Forests (MADREF).[4] Livestock includes about 17 million sheep, 5.3 million goats, 2.7 million cattle, and about 200,000 camels. Although animal husbandry is the sole source of income for only 18 percent of the farmers, it also represents a significant financial reserve for most farmers (FAO 2004).[5]

Demographic growth (1.1 percent in 2009[6]) has contributed to an increased demand for agricultural and livestock products, pushing farmers to overexploit the rangeland resources and to convert productive rangelands to marginal croplands. These factors, combined with the country's natural fragility of soils, are believed to have contributed to the degradation of rangeland, cropland, and forestland in Morocco (Merzouk 1988; Benzyane, Blerot, and Giot 1999; Ouassou, Amziane, and Lajouad 2006).[7]

Methodology

Estimating the magnitude of land degradation and its consequences is difficult. In theory, assessing on-site effects is straightforward—for example, by using the production function approach. In practice, however, the assessment is empirically difficult because of the lack of appropriate data about the yield impact of degradation. Moreover, the site specificity of soil characteristics limits the applicability of data collected in one location to the analysis of problems at another.

Efforts to value off-site effects are also constrained by insufficient data, mainly because of the unclear cause-and-effect relationships in both space and time (Walling 1988). This chapter focuses only on the on-site effects of land degradation on cropland and rangeland in Morocco and refers the results to the year 2000.

Valuing the Cost of Cropland Degradation

Land degradation is defined as "the temporary or permanent reduction in the productive capacity of land as a result of human action" (FAO 2000). The cost of cropland degradation is estimated in three steps:

1. Estimating the share of degraded land in total cropping area
2. Estimating the impact of land degradation on crop productivity
3. Assessing the cost of degraded cropland

Estimating the share of degraded land in total cropping area. FAO (2000) estimates the share of degraded land in total cropping area based on the analysis of the Global Assessment of Soil Degradation (GLASOD) survey data (Oldeman, Hakkeling, and Sombroek 1990; Oldeman 1994). The GLASOD survey establishes a set of geographical areas (mapping units) that are relatively homogeneous in their physical characteristics. Four *degrees* of land degradation are distinguished in terms of reduction in land productivity:

- *Light,* expressed through a low reduction in productivity
- *Moderate,* expressed through a considerable reduction in productivity
- *Strong,* when biological functions of soil are considerably destroyed and there is no potential for rehabilitation and use
- *Extreme,* when biological functions of soil are considerably destroyed and lands cannot recover

In addition, the *spatial extent* of land degradation reflects the share of degraded land in one geographical area (mapping unit). This share is classified within the following five ranges: 0–5 percent, 5–10 percent, 10–25 percent, 25–50 percent, and 50–100 percent.

Twenty combinations of *degree* and *spatial extent* of degradation are therefore possible. Those combinations fall within four *classes* of degradation severity: light, moderate, severe, and very severe. Figure 5.1 shows, for example, that a *severely degraded area* can mean that (a) 10–25 percent of land is subject to strong degradation; (b) 25–50 percent of land is subject to moderate degradation; or (c) 50–100 percent is subject to

Figure 5.1   Classes of Land Degradation Severity

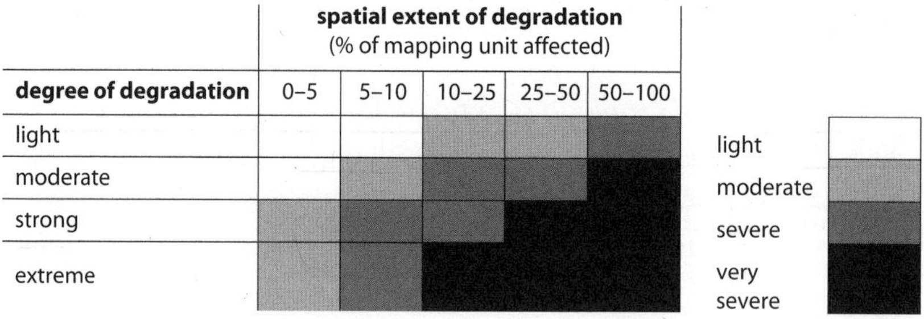

Source: FAO 2000.
Note: The class of degradation severity (light, moderate, severe, or very severe)—as indicated by the shaded areas and keys to the right of the figure—is determined by a combination of the degree and spatial extent of land degradation as defined by the GLASOD survey data.

light degradation. Similarly, a *very severely degraded area* can mean that (a) 10–25 percent of land is subject to extreme degradation; (b) 25–50 percent of land is subject to strong degradation; or (c) 50–100 percent of land is subject to moderate degradation.

Estimating the impact of land degradation on crop productivity. This step estimates the value of lost crop production due to decreased productivity caused by land degradation. It is based on the change in productivity method. Because most agricultural land is cultivated with cereals, the damage is valued in terms of cereal loss.

Assessing the cost of degraded cropland. The cost of degraded cropland is valued based on the total loss in crop production due to degradation and the local market price of cereals.

Valuing the Cost of Rangeland Degradation

This valuation is based on the change in productivity method. The loss in forage productivity is estimated by using the difference in forage yields between degraded and nondegraded rangelands. The monetary value of lost forage production is based on the total loss of forage and its price in local markets.

Cost of Cropland Degradation

Agricultural activities (such as annual cropping) are often blamed as a major cause of land degradation. These activities, however, are only one contributor to land degradation, others being deforestation, overgrazing, and industrial activities (FAO 2000). This section estimates the annual cost of cropland degradation caused by agricultural activities in Morocco.

Estimating the Share of Degraded Land in Total Cropping Area

An estimated 8.7 million hectares, or 19 percent of Morocco's land (excluding the Saharan provinces), is subject to severe and very severe degradation, as presented in table 5.1 (FAO 2000). FAO also notes that agricultural activities are responsible for 100 percent of the land that is severely and very severely degraded in the country.

The methodological section above explained the possible scenarios defining the "severe" degradation class: (a) 10–25 percent of land is subject to strong degradation; (b) 25–50 percent of land is subject to moderate degradation; or (c) 50–100 percent is subject to light degradation.

Table 5.1   Severity of Cropland Degradation in Morocco

Degradation severity class	Area (ha thousands)	Percentage of land[a] in class
None	2,000	4
Light	4,200	9
Moderate	29,700	67
Severe	6,300	14
Very severe	2,400	5
TOTAL AREA[a]	44,700	

Source: FAO 2000.
Note: The "degradation severity class" is based on the combination of degree and spatial extent of degradation, as displayed in figure 5.1. ha = hectares.
a. Excluding Saharan provinces.

None of the surveys on land degradation in Morocco showed a "strong" or "extreme" degree of land degradation, which would indicate a nonrecoverable loss of the soil's biological function.[8] Thus, this analysis uses only scenarios "b" of moderate and "c" of light degradation classes.

Estimating the Impact of Land Degradation on Crop Productivity

Because most agricultural land is cultivated with cereals,[9] the loss of agricultural yield is estimated in terms of lost cereal productivity due to land degradation. In a study to estimate the cost of land degradation in South Asia, Young (1994, 1998) associates the loss in cereal yields with the severity of land degradation: light degradation corresponds to a 5 percent decrease in cereal yield; moderate degradation, to a 20 percent decrease; and strong degradation, to a 75 percent decrease.[10]

From 1980 to 2000, the average cereal yield for Morocco was 10 quintals per hectare.[11] Based on this number and the estimated yield decreases corresponding to light and moderate degradation, the yield loss would be 0.5 quintals per hectare under the light-degradation scenario and 2 quintals per hectare under the moderate-degradation scenario. An analysis of 1980–2000 cereal production statistics confirms the above estimates of yield losses.[12]

Assessing the Cost of Degraded Cropland

To estimate the cost of degradation, we use the average of the lower and upper bounds of the costs of moderate and light degradation. In 2000, the price was US$24 per quintal of bread wheat, US$27 per quintal of durum

Table 5.2   Estimated Cost of Degraded Cropland

	Light degradation		Moderate degradation	
	Lower bound	Upper bound	Lower bound	Upper bound
Extent of degradation (% of land area)	50	100	25	50
Degraded agricultural land (ha thousands)	4,350	8,700	2,175	4,350
Level of decrease (% drop in yield)	5	5	20	20
Decrease in yield (qx/ha)	0.5	0.5	2.0	2.0
Lost production (qx thousands)	2,175	4,350	4,350	8,700
Lost value (US$ millions)	**52**	**104**	**104**	**209**
Average loss (US$ millions)	**78**		**157**	

Source: Adapted from World Bank 2003.

Note: Estimated yield decreases and production losses refer to cereal crops. ha = hectares, qx = quintals.

wheat, and US$21 per quintal of barley—averaging to US$24 per quintal of cereals.

Although degradation reduces the yields, production costs remain largely the same. Loss of agricultural value added is thus equal to the number of lost quintals, evaluated at the selling price of US$24 per quintal of cereals. Accordingly, as table 5.2 shows, the estimated cost of degradation ranges from US$78 million to US$157 million—averaging to US$117.5 million.[13]

Cost of Rangeland Degradation

Morocco's 65 million hectares of rangelands are the primary source of animal food, providing 30 percent of the overall requirements. However, rangelands are under pressure from climatic factors (for example, drought) as well as from animal-related (overgrazing) and human factors (for example, land clearing for cultivation and wood removal). This section estimates the degradation cost of rangelands with steppe and forest dominance.[14] It excludes Saharan provinces because of the lack of information about forage production in this region.

Table 5.3 presents the distribution of rangelands in Morocco by location and vegetation type. It shows that rangelands with steppe dominance cover 12 million hectares and rangelands with forest dominance extend 5.1 million hectares. The Ministry of Agriculture and Agricultural Development

Table 5.3   Distribution of Rangelands by Region and Vegetation Type
ha millions

Location	Steppe	Forest	Total
Rangelands with steppe dominance (a)	**57.3**	**0.7**	**58.0**
Sahara	46.0	0.0	46.0
Pre-Sahara	5.6	0.1	5.7
Oriental zone	4.7	0.3	5.0
North Atlas	1.0	0.3	1.3
Rangelands with forest dominance (b)	**0.4**	**4.7**	**5.1**
Middle Atlas	0.2	1.0	1.2
High Atlas	0.2	2.0	2.2
Rif	0.0	0.9	0.9
Mamora and Central Plateau	0.0	0.8	0.8
Other rangelands (c)	**0.9**	**0.7**	**1.6**
Total rangelands (a + b + c)	**58.6**	**6.1**	**64.7**

Source: MATUHE 2001.
Note: ha = hectares.

reports a forage productivity of 79 forage units per hectare of rangeland with steppe dominance and 558 forage units per hectare of rangeland with forest dominance (MAMVA 1994, 1995). The ministry also suggests two scenarios of productivity loss due to degradation: a lower bound of 6 percent and an upper bound of 10 percent.

Based on the lower and upper bounds of productivity loss (6 percent and 10 percent, respectively), table 5.4 shows the estimated total loss of forage production: 26–44 million units in rangelands with steppe dominance and 32–54 million units in rangelands with forest dominance. Local market price is about US$21 per quintal of barley, which corresponds to US$0.2 per forage unit. Accordingly, the average damage costs are US$12.5 million (under the 6 percent scenario) and US$20.9 million (under the 10 percent scenario)—averaging to US$16.7 million.

Conclusions

The degradation of cropland and rangeland in Morocco exacts an estimated annual cost of US$91 million to US$178 million, or an average of US$134 million—0.4 percent of GDP in 2000. Cropland degradation accounts for 88 percent of this cost, and rangeland degradation for only 12 percent, the latter almost equally divided between areas with forest and steppe dominance, as figure 5.2 depicts. These results do not capture several problems, such as the impact of salinity on irrigated soil. Therefore,

Table 5.4   Estimated Loss of Forage Production and Damage Costs to Rangelands

	Lower bound of productivity loss (6%)		Upper bound of productivity loss (10%)	
	Rangelands with steppe dominance	Rangelands with forest dominance	Rangelands with steppe dominance	Rangelands with forest dominance
Rangeland (ha thousands)	12,000	5,100	12,000	5,100
Degraded area (% of rangeland area)	46	19	46	19
Forage productivity (FU/ha/year)	79	558	79	558
Yield loss due to degradation (%)	6	6	10	10
Total loss due to degradation (FU thousands)	26,170	32,440	43,600	54,070
Cost of forage (US$/FU)	0.2	0.2	0.2	0.2
Damage cost due to degradation (US$ millions)	5.6	6.9	9.3	11.5
Average damage cost per scenario (US$ millions)	12.5		20.9	
Average damage cost (US$ millions)		**16.7**		

Source: Authors' calculations based on the information reported in the main text.
Note: FU = forage unit, ha = hectare. One FU corresponds approximately to 1 kilogram of barley in terms of nutritional content.

Figure 5.2   Annual Cost of Cropland and Rangeland Degradation, by Category

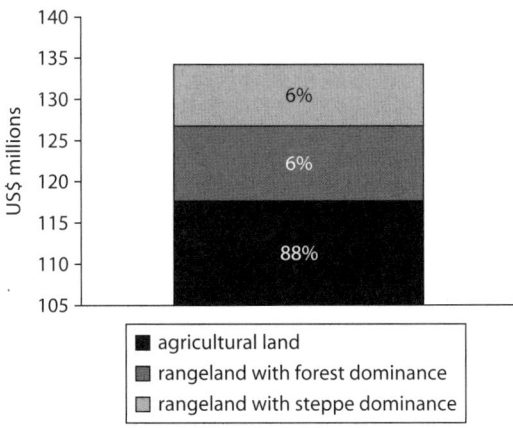

Source: Authors' calculations.

the above estimate most likely underestimates the total impact of land degradation.

If one considers a total population of 30.4 million in Morocco,[15] the damage corresponds to about US$4 per capita in 2000—a conservative figure relative to other countries in the region. For example, land degradation in Africa, which loses 75 billion tons of soil per year, costs approximately US$70 per capita, or US$400 billion per year (Lal 1998).

In addition, the estimate for Morocco corresponds to about 1.6 percent of agricultural GDP—at the lower end of the estimates for Africa. For example, in Sub-Saharan Africa, annual on-site losses from land degradation range from 1 percent of agricultural GDP in Madagascar, Mali, and South Africa to 8 percent in Zimbabwe (Bojo 1996). Three different studies from Ethiopia, however, have estimated annual losses of less than 1 percent (FAO 1986), 4 percent (Bojo and Cassells 1995), and 5 percent of agricultural GDP (Sutcliffe 1993). The varying results attest to the weakness of the data and the dependence on the assumptions made (Pagiola 2002).

The total COED in Morocco is estimated at US$12 billion, or 3.7 percent of GDP in 2000 (World Bank 2003). As figure 5.3 shows, cropland and rangeland degradation account for about 11 percent of the total COED—considerably more than the share attributable to deforestation. In contrast, the costs of air and water degradation are much greater than those related to cropland and rangeland degradation, primarily because of the substantial health impacts of air and water pollution.

Figure 5.3   Shares of Environmental Degradation Cost in Morocco, by Category

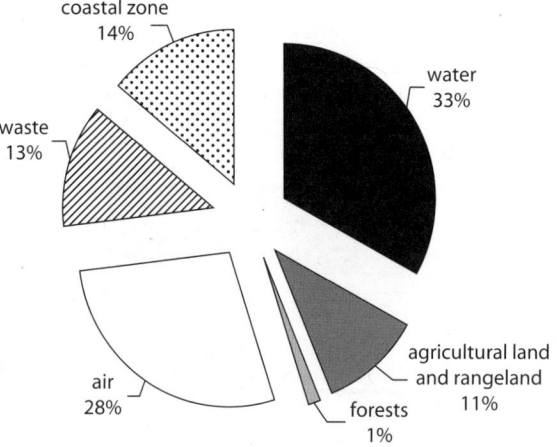

Sources: World Bank 2003; authors' calculations.
Note: Percentages represent the estimated share of the annual US$12 billion COED in Morocco attributable to each environmental degradation category.

Notes

Maria Sarraf is a Senior Environmental Economist at the World Bank, South Asia Region, MC10-1019 1818 H Street, NW Washington, DC 20433, USA.

Abdeljaouad Jorio is former Professor of Economics at Ecole Nationale d'Administration in Rabat and consultant in sustainable development, 22 Rue Zerhoun, Souissi, Rabat 10000, Maroc.

1. Desert and arid land occupy 78 percent of the total land, and semiarid land covers 15 percent of the total land.

2. The value added of agriculture is US$1,170 per hectare of irrigated land versus US$805 per hectare of rain-fed land (FAO 2005).

3. The study also shows that other oil-poor countries of the region can be considerably more vulnerable than Morocco—for example, Djibouti, Jordan, Lebanon, and the Republic of Yemen—because they are highly dependent on cereal imports.

4. In contrast, FAO (2004) provides a figure of 53 million hectares. This chapter uses the MADREF figure because it is considered more reliable.

5. It is particularly important to those farmers with difficult access to agricultural loans.

6. https://www.cia.gov/library/publications/the-world-factbook/geos/mo.html.

7. With an average soil loss of more than 2,000 tons per square kilometer per year in the Rif region, Morocco belongs to the group of countries exhibiting the highest erosion rate (Merzouk 1988; MADREF 2001); forests decline at a rate of 31,000 hectares per year; and about 8.3 million hectares of rangelands are heavily degraded, mainly those in the eastern regions of Morocco, the Souss, the pre-Sahara, and the Sahara (Ouassou, Amziane, and Lajouad 2006).

8. This does not mean that severe degradation does not exist. For example, a FAO study in Morocco shows that since 1975, of 22.7 million hectares of large watersheds, only 8.2 million hectares are suitable for cultivation, of which 50 percent (4.1 million hectares) require urgent soil conservation measures. The remaining 14.5 million hectares should not be cultivated at all. Another survey found that 2.1 million hectares of agricultural land suffer from water erosion (MAMVA 1996). This surface area may have been underestimated, however, because the authors studied only priority sites that required urgent intervention. The actual affected area is likely far greater than 2.1 million hectares, and this area therefore faces "severe" to "very severe" degradation.

9. Cereals cover 5.2 million hectares, or 60 percent of agricultural land, if irrigated land is excluded (World Bank 2001).

10. Similarly, Lal (1995) estimated that soil erosion decreases the productivity of cultivated land in Africa by anywhere between 2 and 40 percent. Also see

Pagiola (1997) for a comprehensive analysis of the environmental problems caused by Moroccan agricultural practices.

11. Government cereal statistics are from MADREF (2001).

12. In addition, during this period, the average production increased by 0.52 percent because of an increased cultivated area, not because of an increased yield.

13. A paper estimating the long-run economic effects of erosion on wheat production concludes that erosion *per se* is unlikely to lead to dramatic deteriorations in the wheat production of the semiarid Morocco; only on steep slopes does erosion appear to be a threat under plausible erosion rates (Pagiola and Bendaoud 1994). Because the paper's results are site-specific, we cannot extrapolate them to the whole country.

14. In this chapter, rangelands with steppe dominance are defined as rangelands where steppe is the main vegetation cover. Similarly, rangelands with forest dominance are the rangelands where forest is the main vegetation cover.

15. See World Bank Development Data Platform, http://data.worldbank.org.

References

Benzyane, M., P. Blerot, and P. Giot. 1999. *Le Grand Livre de la Forêt Marocaine.* Belgium: Editions Mardaga.

Bojo, J. 1996. "The Costs of Land Degradation in Sub-Saharan Africa." *Ecological Economics* 16 (2): 161–73.

Bojo, J., and D. Cassells. 1995. "Land Degradation and Rehabilitation in Ethiopia: A Reassessment." AFTES Working Paper 17, Africa Technical Department, World Bank, Washington, DC.

FAO (Food and Agriculture Organization of the United Nations). 1986. "Highlands Reclamation Study, Ethiopia." Internal report FAO-AG— UTF/ETH/037/ETH, Agriculture Department, FAO, Rome.

———. 2000. "Land Resources Potential and Constraints at Regional and Country Level." World Soil Resources Report 90, FAO, Rome.

———. 2004. "Morocco." Country Pasture/Forage Resources Profile, FAO, Rome. http://www.fao.org/ag/AGP/AGPC/doc/Counprof/Morocco/morocco.htm.

———. 2005. "Irrigation in Africa in Figures: AQUASTAT survey." Water Report 29 series, FAO, Rome. ftp://ftp.fao.org/agl/aglw/docs/wr29_eng_including_countries.pdf.

Lal, R. 1995. "Erosion–Crop Productivity Relationships for Soils of Africa." *Soil Science Society of America Journal* 59 (3): 661–67.

———. 1998. "Soil Erosion Impact on Agronomic Productivity and Environmental Quality." *Critical Reviews in Plant Sciences* 17 (4): 319–464.

MADREF (Ministry of Agriculture, Rural Development, Waters and Forests). 2001. "Statistiques ceréalières." Rabat: MADREF.

MAMVA (Ministry of Agriculture and Agricultural Development). 1994. "Plan National d'Aménagement des Bassins Versants." Phase II, vol. 1, MAMVA, Rabat.

————. 1995. "Plan National d'Aménagement des Bassins Versants: Priorités régionales." Phase II, Synthesis Report, MAMVA, Rabat.

————. 1996. *Résultats de l'enquête nationale sur les terres agricoles soumises à l'érosion*. Rabat: MAMVA.

MATUHE (Ministry of Land Planning, Urban Development, Housing and Environment). 2001. *Rapport sur l'Etat de l'Environnement du Maroc*. Rabat: MATUHE.

Merzouk, A. 1988. "L'érosion hydrique des sols déprécie leur productivité." *Revue des naturels enseignants* 55 (1987–88): 56–65.

Oldeman, L. R. 1994. "The Global Extent of Soil Degradation." In *Soil Resilience and Sustainable Use*, ed. D. J. Greenland and I. Szabolcs, 99–118. Wallingford, Oxfordshire, U.K.: CABI Publishing.

Oldeman, L. R., R. T. A. Hakkeling, and W. G. Sombroek. 1990. *World Map of the Status of Human-Induced Soil Degradation: An Explanatory Note*. Wagenigen, Netherlands: International Soil Reference and Information Centre and United Nations Environment Programme (UNEP).

Ouassou, A., T. H. Amziane, and L. Lajouad. 2006. "State of Natural Resources Degradation in Morocco and Plan of Action for Desertification and Drought Control." In *Desertification in the Mediterranean Region: A Security Issue*, ed. W. G. Kepner, J. L. Rubio, D. A. Mouat, and F. Pedrazzini, 251–68. Dordrecht, Netherlands: Springer.

Pagiola, S. 1997. "Environmental Problems in Moroccan Agriculture." Morocco Environmental Review and Action Strategy, World Bank, Washington, DC.

————. 2002. "Economics of Soil Management in Developing Countries." In *Encyclopedia of Soil Science*, ed. R. Lal, 378–81. Boca Raton, FL: Taylor & Francis.

Pagiola, S., and M. Bendaoud. 1994. "Long-Run Economic Effects of Erosion on Wheat Production in a Semi-Arid Region of Morocco: A Simulation Analysis." Agricultural Economics Staff Working Paper 95–12, World Bank, Washington, DC.

Sutcliffe, J. P. 1993. *Economic Assessment of Land Degradation in Ethiopian Highlands: A Case Study*. Addis Ababa: National Conservation Strategy Secretariat, Ministry of Planning and Economic Development, Government of Ethiopia.

Walling, D. E. 1988. "Erosion and Sediment Yield Research: Some Recent Perspectives." *Journal of Hydrology* 100 (1–3): 113–41.

World Bank. 2001. *World Development Indicators*. Washington, DC: World Bank.

————. 2003. "Kingdom of Morocco: Cost Assessment of Environmental Degradation." Report 25992-MOR, Middle East and North Africa, World Bank, Washington, DC.

World Bank, FAO (Food and Agriculture Organization of the United Nations), and IFAD (International Fund for Agricultural Development). 2009. *Improving Food Security in Arab Countries*. Washington, DC: World Bank, FAO, and IFAD.

Young, A. 1994. "Land Degradation in South Asia: Its Severity, Causes, and Effects upon the People." World Soil Resources Report 78, FAO, Rome.

————. 1998. *Land Resources: Now and for the Future*. Cambridge, U.K.: Cambridge University Press.

CHAPTER 6

Oil Spill and Waste due to Conflict: The Case of Lebanon

Maria Sarraf, Lelia Croitoru, Mutasem El Fadel, Karim El-Jisr, Erkki Ikäheimo, Erich Gundlach, and Samia Al-Duaij

The 34-day hostilities in Lebanon started on July 12, 2006, and continued until August 14, 2006, when the ceasefire went into force. It killed close to 1,200 people, left more than 4,400 injured, displaced more than a quarter of the population, and severely damaged the country's infrastructure (GoL 2006a). Beyond these tragic impacts, the conflict devastated the country's fragile environment. The bombing of a power plant in Jiyeh caused 12,000 to 15,000 tons of oil to spill into the Mediterranean Sea. The widespread fires and oil burning deteriorated air quality, especially in southern Beirut. These damages significantly affected the country's economy, environment, and public health.

Several damage assessments were carried out in the aftermath of the hostilities:

- The Government of Lebanon (2006a) estimated the direct damages to infrastructure and economic sectors.[1]
- The World Bank (2006) assessed the impact on the country's major economic and social sectors.
- The Food and Agriculture Organization (2006) focused on the physical damages and income losses to agriculture, fisheries, and forestry.

- The European Commission (2006a) quantified the direct damages to public infrastructure.

None of these studies focused on estimating the damages to the environment. As of April 2007, only the United Nations Development Program (UNDP) and the United Nations Environment Program (UNEP) had carried out environmental assessments of the damage (UNDP 2007; UNEP 2007). These studies illuminated several important aspects of the environmental degradation; however, none of them measured the associated costs in monetary terms.

To bridge this gap, the World Bank (2007) conducted an economic assessment of the environmental damage caused by the hostilities. The next sections summarize the valuations of (a) the impacts of the oil spill on the coast and (b) the impacts of the hostilities on the waste sector—the two most important areas of environmental damage caused by the 2006 conflict in Lebanon. It is important to note that this analysis was undertaken between October 2006 and April 2007. Several changes may have occurred since then, entailing potential changes in the estimated damage costs.

The Oil Spill

On July 13 and 15, 2006, bombs hit the storage tanks of the Jiyeh power utility, located 30 kilometers south of Beirut—storage tanks that contained approximately 44,000 tons[2] of stored intermediate fuel oil (IFO).[3] As a result, about 12,000 to 15,000 tons of oil spilled into the Mediterranean Sea, and the rest burned, according to communications with the Lebanese Ministry of Environment (MOE) in 2007. Photo 6.1 shows aerial views of the Jiyeh station's tank area before and after the bombings.

Photo 6.1   Jiyeh Electrical Power Station and Tanks

a. Before the July 2006 bombings	**b. After the July 2006 bombings**

Source: Google Earth.   *Source:* Lebanon MOE.

Overview of the Oil Spill

The sea currents and winds moved the spilled oil northward and onto the shoreline. The heaviest impacts occurred between the Jiyeh power station and Beirut, between Byblos and Chekka, and on the Palm Islands off the Tripoli shore. By July 29, 2006, the oil reached Syrian Arab Republic waters and affected the shoreline there, as observed at Tartus. Most of the oil remained relatively close to the shoreline, as shown by the MEDSLIK model (in figure 6.1),[4] satellite image analysis, and aerial surveys. Some oil was observed on the bottom of the sea, particularly in areas adjacent to and offshore from the Jiyeh power plant, Beirut, and Byblos.

A similar occurrence was noted in 1991, when the supertanker Haven burned off the shore of Genoa, Italy, and part of its crude oil cargo sank. The main causes of the bottom oil in Lebanon likely included (a) oil burning, creating neutrally buoyant to slightly heavier-than-water material, and (b) oil mats formed when heavy oil mixed with sediment on the bottom. On September 26, 2006, the Italian mission involved in the damage assessment reported a bottom-oil density of about 1.2 grams per cubic centimeter and a 50 percent weight content of sand.

The *Centre de Documentation, de Recherche et d'Expérimentation sur les pollutions accidentelles des eaux* (Centre of Documentation, Research and Experimentation on Accidental Water Pollution, or CEDRE) analyzed several oil samples and found that the spilled IFO 150 had a biodegradability

Figure 6.1   Oil Impact Areas, July 15 to August 2, 2006

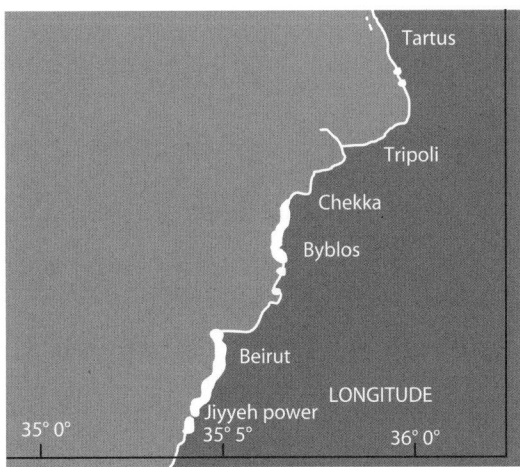

Source: Cyprus Oceanography Center, University of Cyprus (from the MEDSLIK oil-spill model)
Note: The area affected by the oil spill is the light section near the shore.

of about 47 percent, after which nondegradable tars and resins were expected to remain (CEDRE 2006). Although biodegradation was not active on the remaining residues, physical weathering (for example, movement by wave action) was expected to continue removing portions of the remaining residue. The Jiyeh oil had high saturate levels (above 50 percent) and low aromatics (below 28 percent). Because of the oil's low aromatic content, toxicity was relatively low, particularly compared with that of the Prestige spill in 2002 and the Erika spill in 1999, as table 6.1 shows.

Environmental Impacts of the Oil Spill

The Jiyeh oil spill had likely effects on several biodiversity components. The impacts on shoreline biota included direct oiling and smothering of organisms on rock-dominated shorelines. In heavily affected areas, the impacts on sand and gravel beaches may have lasted for weeks or even months. Natural cleansing by wave action, assisted by beach washing through the mitigation effort, aided the recovery on sand and gravel shorelines, as seen in photo 6.2. However, because this oil greatly adheres

Table 6.1   Comparison of Jiyeh Fuel with Other Fuels
% content

Spill incident and fuel type	Saturates	Aromatics	Resins and asphaltenes
Jiyeh (IFO 150)	50	28	22
Erika (heavy fuel no. 6)[a]	23	53	24
Prestige (heavy fuel no. 2 M100)[a]	22	56	22
Testing fuel	20	53	27

Source: CEDRE note by François Merlin to MOE.
Note: IFO 150 = intermediate fuel oil with a viscosity of 150 CentiStokes (cSt) at 50°C.
a. For spill details, visit http://www.cedre.fr.

Photo 6.2   Oil Impact on Major Shoreline Types

a. Byblos oiled gravel beach being cleansed by wave action in active intertidal zone   **b. Enfe oiled steeply sloping rocky shore**

Source: E. Gundlach.

to rocky shorelines, natural cleansing in these areas is slow, and mitigation efforts, such as high-pressure washing, are tedious and time-consuming.

Impacts on subtidal communities. The effects on subtidal communities are most evident where oil mats on the bottom of offshore waters smothered the resident organisms, as observed in the near-shore areas off the Jiyeh power plant and Beirut coast. No observations were made on the effects of the oil spill on bottom communities from chemical toxicity in the water column. Potential injury by smothering occurred only in localized areas where large tar mats were present on the bottom. The Italian mission observed some smothered sponges and corals (madrepores) near the power plant.

Impacts on birds. Shorebirds and marine waterfowl are likely to have been injured because the oil stayed fairly close to the shore, where birds typically feed. From July through October 2006, no dead or heavily oiled birds were reported. However, 92 oiled birds were observed following that period until April 2007 in the Palm Islands Nature Reserve, probably because the oil resurfaced after winter storms on the islands, affecting the winter visitors. Although direct impacts of the spill on birds were relatively minimal, indirect effects are likely to be felt for many years to come.[5] The Government Appointed Committee (GAC) for the management of the reserve observed a major reduction in the number of visiting birds, most likely due to the contamination of habitat and possible loss of access to food.[6]

Impacts on marine reptiles (turtles). The marine turtles found in Lebanese waters include the green turtle (*Chelonia mydas*), the loggerhead turtle (*Caretta caretta*), the Nile soft-shelled turtle (*Trionyx trionguis*), and the leatherback turtle (*Dermochelys coriacea*) (MOA and UNEP 1992; Dimirayak and others 2001; BIL 2003; IUCN 2006). All marine turtles are globally classified as threatened species and are protected by most Mediterranean countries. Loggerheads are an endangered species, and green turtles are critically endangered (IUCN 2006).

The oil spill affected three beaches with turtle nesting: Jbail, Palm Islands Nature Reserve, and Ramlat Al-Baida. It is not known whether other oiled sites, such as Jiyeh and Damour, had active turtle nesting. Turtle areas to the south of Beirut were affected by the direct bombing of the conservation site at Mansouri Village, south of Tyre, rather than by the spill (UNDP 2007).

Impacts on marine mammals. Mammals found in the eastern Mediterranean Sea include dolphins, sperm whales, fin whales, and the rare Mediterranean monk seal. The latter, a critically endangered species, has been occasionally observed on Palm Islands Nature Reserve in the past decade (MOE 2004). While no injuries were reported, the movement of tar balls and other hydrocarbons from the spill with the currents might have affected the already stressed population of remaining seals.

In addition to the above impacts, the media reported a small number of dead fish or fish exhibiting unnatural behavior because of the oil spill. However, indirect effects might have occurred because the food source to fishery was likely reduced as the spill spread. The oil spill also affected large areas of Palm Islands Nature Reserve (Kremer, Pasche, and Kilani 2006), damaging the invertebrate community (crabs, small crustaceans, and mollusks)[7] and plants (golden samphire, sea spurge, glasswort, sea purslane) (Ramadan-Jaradi 2007).

Groundwater contamination through seawater intrusion, which usually occurs in densely fissured zones, is another potential impact of the spill. However, results of the groundwater sample analysis collected from Mina Daliyi in Beirut and the coastal area immediately north of Jiyeh at Saadiyat conducted by UNEP were not conclusive (UNEP 2007).[8]

Methodology

The valuation of damages caused by the oil spill is based on the users' forgone benefits: the difference between the *expected* and *actual* benefits derived from activities on the coast. "Expected benefits" refer to the level of environmental benefits that would have been enjoyed had the oil spill not occurred. "Actual benefits" are those provided after the outbreak of the hostilities.

Valuation challenges. Valuing the impacts of oil spills is particularly challenging. The valuation depends on a wide range of factors, such as the types of goods and services in question, the space and time scale, the sectors affected, the poor quality of available statistics, and the possible existence of irreversible and long-term effects (Chas-Amil and others 2004).

In this light, Grigalunas and others (1986) present the empirical and conceptual problems, including the estimation of nonmarket losses to tourists, the determination of fisheries' losses in the absence of accurate biological data, and the question of whether or how to account for the distribution of costs. Efforts to estimate the impacts of oil spills worldwide have encountered most of these problems, and the Lebanese case is no exception.

Although the subsequent sections address most of the valuation problems, two issues deserve special attention: the time frame of oil spill impacts and the oil spill's particular contribution to the total environmental damage caused by the hostilities.

Time frame of oil spill impacts—Early oil spills have shown that the impact time frames vary widely, depending on many factors such as the type of oil, quantity spilled, type of ecosystem services, and species affected by the oil spill.[9] Estimating the time frame of damage becomes even more complicated when unexpected impacts occur at some extended time after the event.[10] In Lebanon, no precise information is available about the time frame of the Jiyeh spill's effects.[11] Two field visits to the Lebanese coast in October 2006 and March 2007, however, indicated the following:

- The strongest impacts of the oil spill on coastal activities covered the period between the August 14 ceasefire and the end of 2006.
- In 2007, the visual effects of the oil pollution would significantly subside if restoration and cleanup continued.[12]
- In 2008 and after, the oil spill may still have site-level impacts, either because the environment was not appropriately cleaned or because of lingering perceptions that potentially negative health effects persist.

Based on the above, this report adopts a *three-year time frame for the analysis (2006–08), during which losses are assumed to subside gradually*.[13]

Oil spill contribution to the damage—Because the oil spill is a direct consequence of the hostilities themselves, it is difficult to single out its contribution to the overall damage.[14] McCay and others (2004) estimate that the potential impact of oil spills on natural resources may range from as little as 2 percent to as much as 50 percent of the total socioeconomic, environmental, and response-related costs, depending on the type of oil, volume percentile, and other characteristics.

Given the inconclusive information related to other oil spills and the timing of the analysis so shortly after the event, this estimation relies on several assumptions, as illustrated in table 6.2. Because more data were available for 2006 than for successive years (for example, expected income or forgone benefits from coastal activities), the assumptions for 2006 rely on the existing baseline information and, consequently, vary

Table 6.2   Oil Spill Damages to Selected Coastal Activities
% of expected income

	2006		2007	2008
Affected site or activity	*July–Aug.[a]*	*Sept.–Dec.*	*Jan.–Dec.*	*Jan.–Dec.*
Commercial fishing	0	50[b]	5–10	0–5
Shoreside fishing	0	50[b]	5–10	0–5
Hotels	0	10–20	5–10	0–5
Byblos World Heritage Site	0	25–50	5–10	0–5
Beach resorts and chalets[c]	0	25–50	5–10	0–5
Palm Islands Nature Reserve	0	75–100	5–10	0–5
Restaurants	0	75–100	5–10	0–5
Marinas' sport services	0	75–100	5–10	0–5

Source: 2007 interviews with local representatives of each activity.
Note: "Expected income" is the income that would have been expected had the oil spill not occurred.
a. The significant July–August impact of the oil spill was overshadowed by that of the hostilities, which completely halted coastal recreational activities. Thus, the hostilities are assumed to have caused all July–August losses.
b. Percentage of forgone income instead of expected income, based on interviews with fishermen.
c. Because beach and seawater characteristics directly influence recreational services at these sites, the oil spill is assumed to equally affect the activities at beach resorts, chalets, and public beaches in each of the three years.

from one activity to another. In the absence of accurate information, the cleanup efforts are assumed to cause a gradual decline of the oil spill impacts in future years—5–10 percent of expected income in 2007 and 0–5 percent in 2008 for all coastal activities. These assumptions are based on local interviews with representatives of each activity concerned (for example, hotels and restaurants) at the time of valuation. Because no precise information was provided, the numerical ranges take the uncertain impact level into account.

Cost of Degradation due to the Oil Spill

The valuation of the cost of degradation is based on the assumptions presented in the methodology section above. For each activity, the total damage reflects the present value of forgone benefits during 2006–08, uses a 4 percent discount rate, and refers to 2006 as a base year.

Hotels and furnished apartments. The Syndicate of Hotel Owners lists 337 licensed hotels in Lebanon, of which 54 are on the coast and include about 3,500 rooms.[15] The average daily income is US$150 per room.[16] In addition, the coast hosts about 97 furnished apartment buildings,[17] totaling 2,800 units, at an average price of US$220 per night.[18] Consequently, the daily income of fully booked coastal hotels and furnished apartments totals about US$1.1 million. Based on peacetime occupancy rates, the expected income is estimated at US$313.2 million per year.

The oil spill reduced the occupancy rate of hotels and furnished apartments along the coast. In 2006, this reduction was significant, mainly because of the visual signs of oiled beaches and contaminated water. According to interviews with hotel owners, the oil spill reduced the income by 10–20 percent from September through December 2006. In April 2007, both tourism and hotel occupancy rates recovered slightly because of successful cleanup efforts. It is assumed that the hotel industry would be better off and would fully recover in the following years. Accordingly, as the figures in table 6.3 show, the total forgone income due to the oil spill is estimated at *US$23 million to US$60 million.*

Beach resorts. Beach resorts are clubs with daily access to the beach, pools, and other recreational facilities but with no sleeping accommodations.[19] The Lebanese coast hosts about 68 beach resorts.[20] Based on discussions with the Syndicate of Maritime Establishments, there are about 500 daily visitors per beach resort during peak season (May–August) and about 300 daily visitors during the rest of the season (September). If one considers that visitor spending averages US$20 per day,[21] the resorts' expected peacetime income is estimated at about US$55.4 million per year.

Table 6.3   Forgone Coastal Hotel and Apartment Income due to Oil Spill

	Minimum	Maximum	Assumptions
Expected income/day	1.1	1.1	(3,500 rooms x $150) + (2,800 apartments x $220)
Expected income			
Sept.–Dec. 2006	87.0	87.0	Based on expected hotel occupancy of 50% of full capacity in winter (Nov.–Feb.) 75% in spring (Mar.–Apr.) 100% in summer (June–Aug.) 75% in fall (Sept.–Oct.)
Jan.–Dec. 2007	313.2	313.2	
Jan.–Dec. 2008	313.2	313.2	
Forgone income			
2006	8.7	17.4	10–20% of the expected income
2007	15.7	31.3	5–10% of the expected income
2008	0.0	15.7	0–5% of the expected income
PV of forgone income	22.8	59.6	

Source: Authors' estimates and interviews with hotel owners in April 2007.
Note: "Expected income" is the income that would have been expected had the oil spill not occurred.
PV = present value.

The hostilities and the oil spill heavily affected the activities on beach resorts. The Syndicate of Maritime Establishments reported about 60 visitors per beach in September 2006—only 20 percent of the peace-time average. In other words, the hostilities and the oil spill caused beach resorts' expected income to decline by 80 percent in September 2006. There is no documented information concerning the oil spill's contribu-tion to the forgone income from beach activities. It is reasonable to assume, however, that oil pollution of beaches and water substantially affected the recreational activities of nearby beach resorts, on the order of about 25–50 percent of the expected income in September 2006. As a result, the total 2006–08 forgone income due to the oil spill is between *US$5 million and US$13 million.*

Table 6.4 displays the estimated forgone income of beach resorts, chalets, public beaches, and events that can be attributed to the effects of the oil spill.

Table 6.4   Forgone Beach Resort, Chalet, Public Beach, and Event Income due to Oil Spill

US$ millions

	Minimum	Maximum	Explanatory notes
Beach resorts			
Expected income			
2006 (Sept., 30 days)	11.9	11.9	68 (resorts) x 300 (visitors) x 30 days x $20/day
2007 (May 15–Sept., 108 days)	55.4	55.4	68 x [(500 x 47 days) + (300 x 61 days)] x $20/day
2008 (May 15–Sept., 108 days)	55.4	55.4	68 x [(500 x 47 days) + (300 x 61 days)] x $20/day
Forgone income due to spill			*% of expected income*
2006	2.9	6.0	25–50
2007	2.8	5.5	5–10
2008	0.0	2.8	0–5
PV of forgone beach resort income	**5.4**	**13.3**	
Chalets			
Expected income			
2006 (Sept.–Oct., 2 months)	10.0	10.0	$5 million x 2 months
2007 (May–Oct., 6 months)	30.0	30.0	$5 million x 6 months
2008 (May–Oct., 6 months)	30.0	30.0	$5 million x 6 months

(continued)

**Table 6.4   Forgone Beach Resort, Chalet, Public Beach, and Event Income
due to Oil Spill** (continued)
US$ millions

	Minimum	Maximum	Explanatory notes
Forgone income due to spill			% of expected income
2006	2.5	5.0	25–50
2007	1.5	3.0	5–10
2008	0.0	1.5	0–5
PV of forgone chalet income	**3.8**	**8.9**	
Public beaches			
Expected income			
2006 (Sept.)	2.6	2.6	$2.6 million x 1 month
2007 (July–Sept.)	7.8	7.8	$2.6 million x 3 months
2008 (July–Sept.)	7.8	7.8	$2.6 million x 3 months
Forgone income due to spill			% of expected income
2006	0.6	1.3	25–50
2007	0.1	0.2	5–10
2008	0.0	0.1	0–5
PV of forgone public beach income	**0.7**	**1.5**	
Events			
Expected income			
2007 (May–Oct.)	71.4	80.3	Equivalent to 6,000–6,700 events per season
2008 (May–Oct.)	71.4	80.3	Equivalent to 6,000–6,700 events per season
Forgone income due to spill			% of expected income
2006	0	0	0
2007	3.6	8.0	5–10
2008	0.0	4.0	0–5
PV of forgone event income	**3.3**	**11.0**	
PV of forgone beach resort, chalet, public beach, and event income	**13.2**	**34.8**	

Source: Authors' calculations.
Note: "Expected income" is the income that would have been expected had the oil spill not occurred.
PV = present value.

Chalets. Chalet complexes are clubs that include privately owned chalets that can be rented on a seasonal basis.[22] The Lebanese coast hosts 25 chalet complexes north of Jiyeh.[23] The high season for renting chalets covers six months, from May through October. On average, each chalet complex has 200 chalets, which can be rented for about US$1,000 per month.[24] Thus, the income from renting chalets averages US$5 million per month, or US$30 million per year. Assuming that the oil spill contributes to the income decline as it did in the case of beach resorts, the total forgone income to chalets is about *US$4 million to US$9 million.*

Public beaches. These beaches are owned by the state and provide free public access. Based on discussions with the Syndicate of Professional Divers and the nongovernmental organization (NGO) Cedars for Care, there are about 15 public beaches in Lebanon, extending over 10–12 kilometers. The peak season covers three months, from July to September. Most likely, the oil spill affected only the beaches north of Jiyeh—namely, Aabdeh, Tripoli, Batroun, Jbail, Tabarja, Jounieh, Ramlet el Baida, St. Simon, and Rmeileh.

Ramlet el Baida is by far the largest and the most frequented public beach in Lebanon, receiving about 4,600 daily visitors.[25] If one assumes that each of the other public beaches receives only 10 percent of the visitors to Ramlet el Baida, this corresponds to about 460 daily visitors per beach. Accordingly, the average number of visitors to the public beaches affected by the spill is about 8,400 per day.

Because public beach entry is free, the individual benefit is assumed to be about half of that enjoyed by visitors to beach resorts, that is, US$10 per day.[26] Therefore, the expected benefits from using public beaches during high season amount to US$2.6 million per month,[27] or US$7.8 million per annual season. If one assumes that the oil spill contributes to the decline in public beach benefits in a way similar to its effect on beach resorts, the total forgone income is about *US$0.7 million to US$1.5 million.*

Events. Beach resorts and chalets frequently organize weddings and other social events, from May through October. An interview in October 2006 at the Jannah coastal resort in Damour revealed that social events usually draw about 300 participants and cost US$40 per person. Beach resorts can organize at least four events per week during four months per year, and chalet complexes can arrange at least three events per week during six months of the year. Overall, there are about 6,000 to 6,700 events per season, providing an income of *US$71 million to US$80 million per year.*

Security concerns and damaged infrastructure—and, to a lesser extent, the signs of oil and its potential impacts on health—led to a decline in event income. Therefore, it can be conservatively assumed that the 2006 event income declined due to hostilities rather than the oil spill. If one assumes that in 2007 and 2008 the oil spill reduced the event income in a way similar to its effect on beach resorts, the total forgone event income is about *US$3 million to US$11 million.*

Marinas' sports activities. Marinas offer recreational services such as boating, diving, waterskiing, docking, and maintenance of private boats. The hostilities and the oil spill affected marinas' activities in different ways during different periods. For example, the hostilities and the naval blockade halted the marinas' recreational services until September 8, 2006. Afterward, oil pollution of seawater and equipment and concerns about possible health effects led to a decline in the public use of marinas' recreational services and caused losses to private boats' owners.

Losses to marinas from boat rental and water sports—The Movenpick marina rents leisure boats (for fishing and boating) to hotels and private companies and provides jet skis and diving services to the public. An October 2006 interview indicated that the marina's income was about US$150,000 in 2004 from boat and jet ski rentals. The interview also suggested that three other marinas (Riviera, St. George, and Dbayeh) have the same level of revenue and together represent about 60 percent of all marinas' revenue in Lebanon. Thus, marinas' total revenue is estimated at about US$1 million per year, as shown in table 6.5.

The season for recreational activities covers May through October, peaking in July through August, when about 50 percent of the income occurs. Lacking accurate information, the estimate assumes that the May–June and September–October periods each generate about 25 percent of total annual income. Because recreational activities resumed in September 2006, the expected income for the rest of the year was US$250,000. Based on the assumptions presented in table 6.5, the total

Table 6.5   Forgone Marina Income from Boat Rental and Water Sports due to Oil Spill

US$ millions

	Minimum	*Maximum*	*% of expected income*
Expected income			
2006 (Sept.–Oct.)	250	250	25
2007 (May–Oct.)	1,000	1,000	100
2008 (May–Oct.)	1,000	1,000	100
Forgone income due to spill			
2006 (Sept.–Oct.)	188	250	75–100
2007 (May–Oct.)	50	100	5–10
2008 (May–Oct.)	0	50	0–5
PV of forgone income	**238**	**377**	

Source: Authors' estimates and interview at Movenpick marina in October 2006.
Note: "Expected income" is the income that would have been expected had the oil spill not occurred.
PV = present value.

forgone income from recreational activities at marinas is estimated to be between *US$238,000 and US$377,000.*

Losses to owners of private boats—The oil spill also damaged many private leisure boats docked in marinas and fishing boats docked in ports. The oil damage limited the owners' benefits from using their boats throughout 2006 and imposed additional cleaning costs. The forgone benefit to private owners from not using the boats from September through December 2006 is assumed to equal at least the value of the annual depreciation of the boats and the maintenance costs (cost of upkeep and docking in marinas).[28] The loss due to oiled fishing boats is estimated based on the annual maintenance costs.[29]

Interviews with several marina managers revealed that about 1,775 boats were docked in marinas.[30] Because they belong to marinas north of Jiyeh, theoretically all of them were oiled. In reality, many marinas escaped the oil spill because of their orientations and sea currents. Because the available information does not distinguish between oiled and clean boats, only 50 percent, or 890 boats, are assumed to have been oiled. In addition, observations during cleaning operations show that an additional 20 fishing boats were oiled in Daliyi port alone. Photo 6.3 illustrates fishing boats oiled in Daliyi port.

A boat of average size (6 to 12 meters) has a price of about US$30,000 and a lifetime of 20 years,[31] hence an annual depreciation value of

Photo 6.3   Oil Pollution in the Port of Daliyi

Source: Ministry of Environment of Lebanon.

US$1,500. The cost of upkeep and docking in marinas is conservatively estimated at US$300 per meter per season. Assuming an average boat size of 9 meters, the annual cost of upkeep and docking is about US$2,700 per boat.[32]

The loss to owners of private leisure boats (890 boats) is based on the annual depreciation value and maintenance costs (US$4,200), totaling US$3.7 million. The loss to owners of oiled fishing boats (20 boats) is estimated based on the annual maintenance costs (US$2,700), amounting to US$54,000. Adding the two loss figures, the total damage to private owners of leisure and fishing boats is about *US$3.8 million*. When that estimate is added to the other marina-related loss estimates above, the overall losses to marinas' sports activities are *US$4 million to US$4.2 million*.

Palm Islands Nature Reserve. Palm Islands Nature Reserve is a marine reserve and a Mediterranean Specially Protected Area under the Barcelona Convention (1995) and the Ramsar Convention of Wetlands (1971). Access to the reserve was halted from the start of hostilities until the lift of the naval blockade. The hostilities and oil spill considerably reduced the tourism and associated revenues to local communities (for example, transportation and other services) and affected the area's biodiversity (for example, by oiling birds and turtles).

Loss of recreation—The oil spill played a major role in reducing the number of visitors to Palm Islands Nature Reserve, especially after the end of the naval blockade. The loss in tourism in 2006 is estimated by the difference between the expected number of tourists and actual arrivals.

About 80 percent of tourists to the reserve use its facilities for boat transportation and group excursions to islands. Over the course of a tourist season, there are usually 500 visiting groups of about 15 people per group. The forgone number of individual visitors and groups who use the Palm Islands Nature Reserve facilities is estimated at 20,760 individuals, as shown in table 6.6. The estimated expected annual income from tourist activities is US$72,000, based on a tourist season of about 13 weeks (July–September) and the fees for each recreational activity (table 6.7). Therefore, the total loss in tourism-related income to the reserve due to the oil spill is valued at about *US$15,300 to US$25,900*.

Damage to biodiversity—Around 92 oiled birds from 19 different species were observed in the Palm Islands Nature Reserve, one of which is displayed in photo 6.4.[33]

Table 6.6   Individual Visitors and Groups Using Palm Island Nature Reserve Boats

Visitor type	Number of visitors
Expected visitors in 2006[a]	**22,500**
Individual visitors by Palm Islands Nature Reserve boats[b]	10,500
Groups by Palm Islands Nature Reserve boats[c]	500
Actual visitors in 2006	**1,740**
Individual visitors by Palm Islands Nature Reserve boats[d]	812
Groups by Palm Islands Nature Reserve boats	62
Forgone visitors in 2006[e]	**20,760**
Individual visitors by Palm Islands Nature Reserve boats	9,688
Groups by Palm Islands Nature Reserve boats	438

Source: Authors' estimates.
Note: "Expected visitors" is the number of visitors that would have been expected had the oil spill not occurred. PV = present value.
a. Number varies between 20,000 and 25,000 (MOE statistics).
b. Number represents the difference between the total number of visitors using Palm Islands Nature Reserve boats (80 percent of 22,500) and the number of group visitors using Palm Islands Nature Reserve boats (500 × 15).
c. There are about 500 groups of around 15 visitors per group every year (Communication Jaradi, April 2007).
d. Number represents same ratio of individual visitors to total visitors as used in estimating "expected visitors."
e. Number represents the difference between the expected and the actual number of visitors.

Photo 6.4   Oiled Bird

Source: Italian Task Force.

In addition, three dead loggerhead turtles (*Caretta caretta*) were reported on Palm Islands Nature Reserve.[34] The insufficiency of the data makes it difficult to estimate the damages to loggerhead turtles.[35] We estimate only the damages to birds based on a restoration cost model developed by McCay and others (2004), which relates the cost per bird to the average abundance per unit area:[36]

$$y = 10,260 * e^{-0.0138*x}, \tag{6.1}$$

where x = annual mean abundance (number per square kilometer), y = cost per bird (US$), e = 2.718.

Table 6.7   Palm Island Nature Reserve: Forgone Income due to the Oil Spill
(thousands US$)

Income	Minimum (US$ thousands)	Maximum (US$ thousands)
Expected annual income (13 weeks)[a]	**72.4**	**72.4**
Forgone income due to the oil spill		
2006[b] (3 weeks)	12.5	16.7
2007[c] (13 weeks)	3.6	7.2
2008[d] (13 weeks)	0.0	3.6
PV of forgone income	**15.3**	**25.9**

Source: Authors' estimates.

Note: "Expected income" is the number of visitors and amount of income, respectively, that would have been expected had the oil spill not occurred. PINR = Palm Island Nature Reserve. PV = present value.

a. Expected annual income is based on an average fee of US$62.50 per group, US$4.00 for individual transportation, and US$6.00 for renting chairs and umbrellas (5 percent of visitors rent chairs and umbrellas): (62.5 × 438) + (4 × 9,688) + (0.05 × 6 × 20,760) = 72,355.

b. Range between a minimum of (0.75 × 72,400 × 3/13) and a maximum of (1.0 × 72,400 × 3/13).

c. Range between a minimum of (0.05 × 72,400) and a maximum of (0.1 × 72,400).

d. Range between a minimum of (0.00 × 72,400) and a maximum of (0.05 × 72,400).

Applying this model to the injured species[37] provides a total damage cost estimate of US$48,600. Loureiro and others (2006) indicate that the birds found and collected after an oil spill typically represent only 15–50 percent of all the oil-killed birds. If one assumes the same range for the Lebanon case, the total damage cost associated with the oiled birds falls between *US$97,200 and US$324,000*. The result should be regarded with extreme caution, however, because no other studies valuing these damages were found.

Cost of impact assessment and monitoring—The MOE carried out an impact assessment study of the oil spill on Palm Islands Nature Reserve biodiversity, which cost about US$27,000.[38] In addition, a long-term monitoring program foreseen for the reserve and other ecologically significant sites affected by the spill is estimated to cost about US$1.2 million to US$1.7 million over a period of 7 to 10 years.[39] Thus, the total cost of the impact assessment and monitoring program is about US$1.2 million to US$1.7 million. Only a part of this cost is likely directly related to the oil spill damage, the rest being an expression of the willingness to pay (WTP) for future information.

If one assumes that 50 percent of the total impact assessment and monitoring cost is due to the oil spill damage, this represents about US$600,000 to US$850,000. Added to the above estimates, the overall impact of the oil spill on the Palm Islands Nature Reserve and other ecologically sensitive areas amounts to *US$0.7 million to US$1.2 million.*[40]

Byblos World Heritage Site. Built during Phoenician times, Byblos is considered the oldest inhabited city in the world and is designated a World Heritage site. The oil spill heavily contaminated the harbor, two medieval towers at its entrance, and other ancient ruins below the archaeological tell in Byblos (UNDP 2007). This reduced significantly the number of visitors and threatened the historical value of the ruins.

Loss of recreational-tourist value—Visits to Byblos take place throughout the year and are usually organized both by tour operators and private individuals. According to the Ministry of Tourism, there are 22 tour operators in Lebanon, of which at least 8 organize trips to Byblos.[41] There are about 300 visitors per year, and the fee is US$30 per person if meals are excluded.[42] Accordingly, the tour operators' annual income from organizing visits to Byblos is about US$72,000.

The proximity of Byblos to Beirut (only 40 kilometers away) suggests that more visitors travel there in private cars than with tour operators. Assuming that they number twice as many as those coming with tour operators and that their average spending is US$15 per person,[43] the annual income from individual trips would be about US$72,000. Thus, the annual income from all visits to Byblos is about US$144,000. The associated damage to tourism in Byblos and other historical towns ranges between US$15,300 and US$42,800, as shown in table 6.8.

Loss of historical-cultural value—No studies estimating the losses of historical-cultural value of Lebanese sites were found.[44] Thus, the present valuation relies on the restoration cost method. A September 2006 mission

Table 6.8   Forgone Tourism Income from the Byblos World Heritage Site
US$ thousands

Income	Minimum	Maximum	Assumptions
Expected annual income	144	144	(8 tour operators x 300 visitors x US$30) + (4,800 individual visitors x US$15)
Forgone income due to oil spill			
2006 (Sept.–Dec.)	9.0	24.0	25–50% of expected Sept.–Dec. income
2007 (Jan.–Dec.)	7.2	14.4	5–10% of annual expected income
2008 (Jan.–Dec.)	0.0	7.2	0–5% of annual expected income
PV of forgone income	**15.3**	**42.8**	

Source: Authors' estimates and interviews with tour operators.
Note: "Expected income" is the income that would have been expected had the oil spill not occurred.
PV = present value.

undertaken by a team from the United Nations Educational, Scientific and Cultural Organization (UNESCO) declared that the most serious damages of historical-cultural value resulting from the hostilities concern the World Heritage site in Byblos.

Accordingly, a special procedure to clean the archaeological remains covered by fuel was recommended. If one assumes that the stones were cleaned manually with a specially prepared solution according to the components of the fuel, the total cleanup cost of operations would be US$100,000 (UNDP 2007). This figure is assumed as the minimum bound of the damage caused by the oil spill to archaeological sites. Overall, the estimated damages to Byblos range between *US$115,300 and US$142,800.*

Restaurants. Fish is an important food served in many Lebanese restaurants on the coast. Both the hostilities and the oil spill negatively affected the activity of these restaurants. The hostilities reduced the number of tourists, which decreased fish demand and consumption. The oil spill also contributed to this reduction, mainly because of people's fears that contaminated fish could harm human health.

According to the Syndicate of Restaurant Owners, about 170 restaurants specialize in fish, of which 150 are on the seashore and the rest inland.[45] Based on the same source, the annual revenue of a fish restaurant is in the range of US$200,000 to US$600,000, averaging to US$400,000 per year, or US$33,000 per month. The interviews with the restaurant owners suggested that the oil spill reduced the expected income from September to December 2006 by about 75–100 percent, as shown in table 6.9. Therefore, the total forgone benefits to restaurants due to the oil spill range between *US$19.5 million and US$31.1 million.*

Fishing. In Lebanon, fishing is usually artisanal and small-scale. It supports about 30,000 fishermen (IUCN/Green Line 2006) who catch an average of 8,000 tons of fish per year (FAO 2006). The oil spill caused *direct* losses by oiling boats and gears and *indirect* losses by reducing the demand for fish, either because of actual fish contamination or the perceived ill effects of the fish on health. Below are the estimated impacts of the oil spill on commercial and seashore fishing.

Commercial fishing—FAO (2006) provides information on the fish catch per season and total income from fishing in 2004. The fish catch varies widely across seasons, accounting for 30 percent of the annual catch in

Table 6.9   Income Forgone by Restaurants due to the Oil Spill

	Minimum	Maximum	Assumptions
Total number of fish restaurants	170	170	n.a.
Annual turnover (US$ thousands/restaurant/year)	400	400	n.a.
Monthly turnover (US$ thousands/restaurant/month)	33.3	33.3	n.a.
Expected income Sept.–Dec. 2006 (US$ thousands/restaurant)	133.3	133.3	n.a.
Forgone income due to oil spill			
2006 (US$ millions)	17.0	22.7	75–100% expected income, Sept.–Dec. 2006
2007 (US$ millions)	3.4	6.8	5–10% expected annual income, 2007
2008 (US$ millions)	0	3.4	0–5% expected annual income, 2008
PV of forgone income due to oil spill	**19.5**	**31.1**	

Source: Authors' estimates and interviews at the Syndicate of Restaurant Owners in October 2006 and April 2007.
Notes: "Expected income" is the income that would have been expected had the oil spill not occurred.
PV = present value. n.a. = not applicable.

spring, 42 percent in summer, 22 percent in autumn, and only 8 percent in winter. The annual income from fishing is about US$31 million. Applying the seasonal catch factor in terms of percentage of total catch to total income, the expected fish income from September through December 2006 is estimated at US$7.4 million, as shown in table 6.10.

A University of Balamand survey of about 200 fishermen from North Lebanon found that the hostilities and the oil spill reduced their income by 45 percent.[46] If one extrapolates the estimates to all fishermen on the coast north of Jiyeh and conservatively assumes that only 50 percent of this income drop is attributable to the oil spill, the associated damage cost in 2006 is about US$1.3 million. Consequently, the total losses during the 2006–08 period are estimated at *US$3 million to US$6 million.*

Shoreside fishing—Shoreside fishing is popular in Lebanon for both recreation and consumption. No accurate information on the impacts of the oil spill on recreational fishing is available, except that it reduced fish price and catch. Thus, this report assumes that the oil spill affected recreational fishermen similarly to commercial ones.

In the south of Lebanon, there are about 1,300 anglers,[47] who account for one-third of the total number in the country.[48] As the oil spread from Jiyeh toward the north, it is assumed that it affected the remaining

Table 6.10 Income Forgone to Commercial and Shoreside Fishing due to the Oil Spill
US$ millions

	Commercial fishing			Shoreside fishing		
	Minimum	Maximum	Assumptions	Minimum	Maximum	Assumptions
Expected annual income	31.0	31.0		2.3	2.3	
Expected income (Sept.–Dec.)	7.4	7.4	$31 million x (0.22 + 0.08/3)	0.7	0.7	50% of total forgone loss (50% x 45% x $0.7 million)
Forgone income due to the oil spill						
2006 (Sept.–Dec.)	1.3	1.3	50% of total forgone loss (50% x 45% x $6 million)	0.16	0.16	50% of forgone income (Sept.–Dec.)
2007	1.6	3.1	5–10% of expected annual income	0.12	0.12	5–10% of expected annual income
2008	0.0	1.6	0–5% of expected annual income	0.0	0.12	0–5% of expected annual income
PV of income forgone due to oil spill	**3.0**	**5.9**		**0.26**	**0.47**	

Source: Authors' estimates and FAO 2006.
Note: "Expected income" is the income that would have been expected had the oil spill not occurred. PV = present value.

two-thirds of the total number, or another 2,600 anglers. Based on an average catch of 2 kilograms per day for a minimum of 50 days and an average price of US$4 per kilogram (FAO 2006), the consumption value of the shoreside catch is US$1 million per year. In the absence of information on the recreational value of shoreside fishing in Lebanon, we assume it is similar to that of public beaches (US$10 per day), amounting to a total recreational value of US$1.3 million per year. Overall, the annual value of shoreside fishing is US$2.3 million.

Considering that the fish catch varies seasonally in the same proportion as in commercial fishing, the expected shoreside fish income from September through December 2006 is estimated at about US$0.7 million. The impact of the oil spill on shoreside fishing is valued based on the same percentages adopted for commercial fishing. Accordingly, the present value of the benefits forgone during 2006–08 is in the range of *US$260,000 to US$472,000.*

Overall, the aggregated impact of the oil spill on commercial and shoreside fishing is in the range of *US$3.2 million to US$6.5 million.*

Oil fuel spilled and burned in Jiyeh. Of the total estimated 44,000 tons of IFO 150 stored at the Jiyeh electrical power plant, about 12,000 to 15,000 tons of oil leaked into the sea and the rest burned. At an approximate cost of US$450 per ton, this loss is estimated at US$20 million. In addition, the cost of hiring three floating tankers to replace the burned tanks is estimated at around US$4 million. The maintenance and operation of the floating tankers, transfer of fuel from different plants to Jiyeh power plant, and soil testing at the burned tanks' site is valued at US$15 million.[49] Overall, the direct loss due to the spilling and burning of Jiyeh fuel oil is *US$39.1 million.*

Oil spill cleanup. The expenditures related to the oil spill cleanup include the direct cleanup cost, the cost of treating and transporting the oiled waste, and the cost of monitoring the cleanup operations.

Cleanup cost—Soon after the ceasefire, the MOE estimated the cost of oil spill cleanup in the range of US$137 million to US$205 million (MOE 2007b), based on an average oil spill cleanup cost of US$13,800 per ton (MOE 2006a, 2006b). The cleanup priorities set by the MOE included two phases (PCM 2007):

1. Removal of free-floating mobile oil from the sea and shore and removal of contaminated debris, including sand, pebbles, used equipment, and garbage

2. Cleanup of polluted sites to a higher level of cleanliness, depending on the nature and the environmental and economic sensitivity of the site

The cleanup expenditures as of April 2007 are estimated at US$14.9 million (SDC 2007; UNDP 2007; communication with experts).[50]

Cost of oiled waste—The first cleanup phase (illustrated in photo 6.5) generated about 1,030 cubic meters of liquid waste and 6,250 cubic meters of solid waste (PCM 2007).[51] We estimate the cost of treating and transporting the oiled waste based on the waste management options considered by the MOE (2007c). Accordingly, if the liquid waste is reprocessed at the Zahrani refinery, it would cost about *US$92,000.*

For the solid waste, options to treat low- to medium-contaminated sand are different from those to treat heavily contaminated sand and solid waste. Most of the 6,250 cubic meters is considered hazardous waste because of the toxicity of the fuel (MOE 2007b, 2007c). Lacking more accurate information, it is assumed that 25 percent of the solid waste includes low- to medium-contaminated sand, and 75 percent represents heavily contaminated sand and pebbles.

If one considers that (a) the quantity of low- to medium-contaminated sand will be reused in the cement, construction, or asphalt industries; (b) the unit cost is US$10 per cubic meter;[52] and (c) the transport cost is US$80,000, the total cost of transporting and treating the low- to medium-contaminated sand is estimated at US$96,000. The heavily contaminated sand and pebbles are likely to be shipped under the Basel

Photo 6.5   Containers of Oiled Waste

Source: M. El Sarji.

convention. At a cost of US$10,000 per cubic meter,[53] the total cost to ship and treat the heavily contaminated sand and pebbles would be about US$47 million.

If one adds the costs of transporting and treating all the oiled waste resulting from the first cleanup phase, the total is estimated at *US$47.1 million*.

Cost of monitoring operations—In addition to the estimated cost of oil spill cleanup already spent as of April 2007, more cleanup and monitoring are needed. Some of these costs are estimated based on discussions with MOE. They include the costs related to the second phase of the cleanup and monitoring operations in the Palm Islands Nature Reserve (US$1 million) and the costs of monitoring pilot sites along the Lebanese coast (US$0.5 million), totaling *US$1.5 million* (World Bank 2007).

Overall, the cost of oil spill cleanup, treatment of oiled waste, and monitoring the Lebanese coast is estimated at *US$63.5 million*.

Summary: Oil Spill Damages

The overall damage and cleanup cost due to the oil spill is conservatively estimated at about *US$203 million*, or *1 percent of GDP* in 2006. Table 6.11 and figure 6.2 present the main components of this cost.

Table 6.11   Estimated Costs of Damage and Cleanup due to the Oil Spill
(US$ millions)

Type of cost	Minimum	Maximum	Mean	% of total
Damage costs				
Cost of oil fuel burned	39.1	39.1	39.1	19
Restaurants	19.5	31.1	25.3	12
Hotels and furnished apartments	22.8	59.6	41.2	20
Beach resorts, chalets, public				
beaches, events	13.2	34.8	24.0	12
Marinas' sports activities	4.0	4.2	4.1	2
Commercial fishing	3.0	5.9	4.4	2
Seashore fishing	0.3	0.5	0.4	0
Palm Islands Nature Reserve	0.7	1.2	1.0	1
Byblos World Heritage Site	0.1	0.1	0.1	0
Total damage costs	**102.8**	**176.4**	**139.6**	**69**
Oil spill cleanup				
Expenditures as of April 2007	14.9	14.9	14.9	7
Oiled waste	47.1	49.9	47.4	23
Monitoring expenses	1.5	1.5	1.5	1
Total cleanup costs	**63.5**	**63.5**	**63.5**	**31**
TOTAL COSTS OF OIL SPILL	**166.3**	**239.9**	**203.1**	**100**

Source: Authors' calculations.

Figure 6.2   Annual Degradation Cost Caused by the Oil Spill

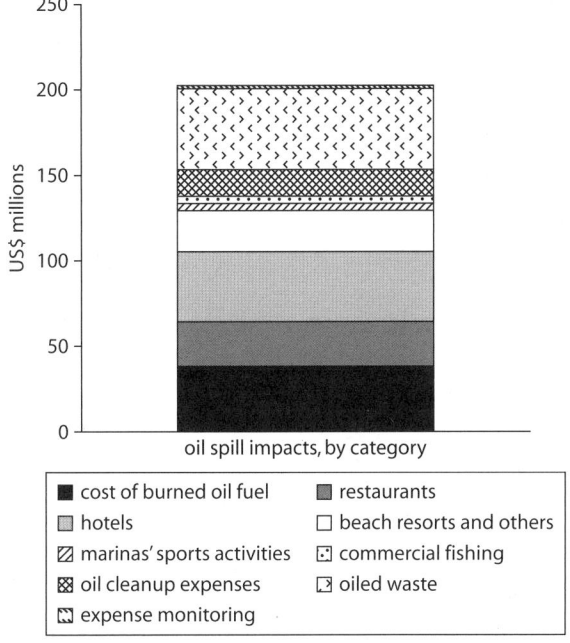

oil spill impacts, by category

- ■ cost of burned oil fuel
- ▣ hotels
- ▨ marinas' sports activities
- ▦ oil cleanup expenses
- ▧ expense monitoring
- ■ restaurants
- □ beach resorts and others
- ⊡ commercial fishing
- ▨ oiled waste

Source: Authors' calculations.
Note: NPV = net present value (t = 25 years, r = 4 percent).

The damage cost accounts for nearly 70 percent of the total oil spill cost. Such a high share is attributable mainly to the costs associated with the burned oil and the losses to hotels, beach resorts, and restaurants. Relatively important damages also occurred to marinas, sports activities, and commercial fishing. The lowest estimates—for damage to the Palm Islands Nature Reserve and Byblos World Heritage Site—are mainly a result of scarce data and do not suggest that they are the least important values.

The cleanup cost accounts for about 30 percent of the total oil spill cost and is dominated by the treatment and shipment of oiled waste, as a result of the expensive shipping procedures under the Basel convention.

It should be noted that the total estimate represents the lower bound of real costs because it does not capture several damage costs, such as the effects on health (for example, skin diseases), on ecosystem services (for example, loss in habitat for spawning), and on marine biodiversity. The estimate also fails to cover the cost of many cleanup operations expected to be performed after April 2007. In addition, for many impacts, the aggregate result tends to reflect the real cost of the oil spill only in part because of the conservative assumptions adopted for valuation. Moreover, the overall estimate and its breakdown by impact should be regarded with much caution because many of the assumptions are subjective for lack of accurate data.

Demolition, Military, and Medical Waste

The hostilities in Lebanon destroyed the country's infrastructure in many areas, leaving enormous amounts of demolition, military, and medical waste. This section focuses on (a) the impacts of demolition waste resulting from the military aggression; (b) the impacts of unexploded ordnances (UXOs), as environmental waste, on people's lives and sources of income; and (c) the increase in medical waste associated with the humanitarian relief effort.

Overview

During the hostilities, the bulk of the military operations concentrated in three areas: the southern suburbs of Beirut, the districts of the South, and the Baalbek El-Hermel region. Destruction of residential units in these areas caused significant quantities of demolition waste. The constituents of typical demolition debris can be grouped into the following categories:

- *Primary inert fractions:* asphalt, brick, cinder block, concrete with rebar or wire mesh, concrete without steel reinforcement, masonite or slate, ceramic tile, glass, dirt or earth, plastic sheet film, plastic pipe, porcelain including bathroom fixtures, ferrous and nonferrous metal, electrical wiring, fiberglass insulation, and plastic buckets or containers.
- *High organic-based fractions:* ceiling tiles; corrugated shipping containers; insulation-treated cellulose; insulation sheathing; pallets, spools, and reels; pressboard or chipboard; roofing materials (such as roofing felt and asphalt shingles); dimensional lumber and shapes (clean); plywood and particle board; and oriented strand board.
- *A range of composite materials* (that may require special handling): carpeting, carpet padding, gypsum wallboard (mainly gypsum with paper backing), electrical fixtures (metal, light tubes or bulbs, and ballasts), electrical switches, rubber hosing and conduits, tires (some with wheels), painted wood, pressure-treated wood, and wood composites.
- *In addition, furniture, electronic appliances, and personal belongings* constitute a considerable portion of the demolition waste resulting from destruction by military activities.

The demolition waste caused damages related to waste hauling, disposal in landfills, road depreciation during transport, traffic delays, and depreciation of land surrounding dump sites. In addition, the demolition waste in ponds and valleys most likely provoked ecosystem damages, such as groundwater pollution.

As of April 2007, there were 864 cluster bomb strike locations in South Lebanon, with an estimated *1 million unexploded cluster munitions* on the ground, contaminating a total of 34 million square meters.[54] Between August 14, 2006, and April 03, 2007, UXOs caused 29 deaths and 195 injuries among residents of South Lebanon (MACCSL 2007). The UXOs also prevented access and exploitation of agricultural land, rangeland, and forests.

With the rise in casualties and hospital bed occupancy numbers, generation of medical waste also increased significantly. The hostilities caused 1,200 deaths and 4,400 injuries (GoL 2006a), generating an estimated 200 to 250 tons of medical waste (UNDP 2007), with impacts on the treatment, transport, and disposal costs.

Methodology

Valuation of the impacts of waste generated during the hostilities refers to 2006 as a base year. It should be noted that the lack of information concerning the extent of the impact sometimes prevented valuation, as in the case of groundwater pollution by demolition waste. Safety concerns also made it difficult to arrive at accurate estimates—for example, of the impacts of UXOs on agricultural land. Thus, the estimates tend to undervalue the real damage.

Impacts of demolition waste. The additional costs or damages generated by waste loading, transport, road depreciation, traffic delays, and disposal are based on cost-based methods:[55]

- *Damages caused by loading, hauling and transport* are valued based on the cost of these activities on the market.
- *Road depreciation* is valued based on the cost of refurbishing the roads damaged by demolition waste transport.
- *Traffic delays* are valued based on the cost of fuel and the opportunity cost of the time lost in traffic due to demolition waste.
- *The cost of land for waste disposal* is valued based on the market price of land near the damaged sites.

Impacts of military waste. The cost of human deaths and injuries caused by military waste is estimated in terms of disability-adjusted life years (DALYs).[56] The DALY provides a common measure of disease burden for illnesses and premature mortality.

To estimate the value of 1 DALY, this chapter uses two approaches. The human capital approach (HCA) estimates the indirect cost of productivity loss through the value of an individual's future earnings (Kirch 2008). Accordingly, 1 DALY corresponds to one person's average contribution to production, namely the GDP per capita. This method provides a lower bound for the loss of 1 DALY.

The second approach, the value of a statistical life (VSL), measures the WTP to avoid death. This estimate is based on observing individual behavior when trading off health risks and money (Johansson 2006). The VSL is calculated by dividing the marginal WTP to reduce the risk of death by the size of the risk reduction. By this measurement, the value of 1 DALY corresponds to the VSL divided by the number of discounted average years of life lost because of an adult's death (World Bank 2005). The VSL method provides an upper bound of health damages.

The impact of military waste on access to agricultural lands is based on the value of lost income due to the scattered UXOs. Lacking more accurate information, it is assumed that the limited access would have a significant impact on farmers for at least two years. In addition to the above costs, valuation includes the costs of demining carried out by the government.

Impacts of medical waste. These impacts are estimated based on the costs of handling the medical waste generated during the hostilities, which include the costs of treatment, transport, and disposal.

Damage Costs due to Demolition Waste

To assess the environmental damage associated with demolition waste, the analysis considers the impact of the actual handling of the waste after the ceasefire. The following subsections estimate the generated quantities of demolition waste and the associated costs of hauling and transport, road depreciation, traffic delays, and the cost of land for waste disposal.

Generated quantities. Government agencies and international organizations conducted field visits to assess the extent of the physical damage in Beirut's southern suburbs, the South, and Baalbek El-Hermel. These efforts resulted in different estimates of the generated quantities of demolition waste. This subsection presents the available information as of April 2007 and adopts the most recent government estimate as an input to the analysis.

Beirut's southern suburbs—UNDP reported that about 150 residential buildings were destroyed, each building containing an average of 30 units, for a total of 4,500 units (UNDP 2006). Other buildings had been damaged or partially demolished.

A joint effort between the Municipality of Haret Hreik and the Department of Architecture and Graphic Design at the American University of Beirut assessed 102 completely demolished buildings, 28 partially blasted buildings, and 70 damaged buildings in August 2006, as the set of images in photo 6.6 depict.

A rapid preliminary damage assessment by the European Commission's Joint Research Center and the European Union Satellite Center indicated that 326 residential buildings were either damaged or destroyed in the southern suburbs, of which 269 were in Haret Hreik (EC 2006b). The Order of Engineering in Beirut assessed 200 destroyed buildings and an additional 100 inhabitable buildings (GoL 2006b).

South and Baalbek El-Hermel—The study team conducted a field survey in October and November 2006 during the preparation of this study. Interviews with heads of municipalities revealed that more than 8,790 housing units were demolished in the South. Most are concentrated in the Cazas of Marjeyoun, Nabatieh, Bent Jbeil, and Tyre, with relatively less destruction in the Cazas of Hasbaya and Saida.

In Baalbek El-Hermel, the field visit revealed that the hostilities affected more than 6,000 housing units, of which 375 were destroyed, 400 badly damaged, and the rest severely to lightly damaged. Based on this information, the authors' own estimate of the quantity of demolition waste generated varies between 2 million to 3.7 million cubic meters, as shown in table 6.12.

A UNDP-sponsored initiative to assess the environmental damage of the July hostilities estimated the total volume of rubble resulting from destruction in the range of 2.5 million to 3 million cubic meters (UNDP 2007).

Based on an August 2007 communication with the office of the President of the Council of Ministers (PCM), the latest figures indicate 11,140 housing units destroyed, 1,249 housing units partially destroyed, and 81,000 housing units lightly damaged in the South and the Baalback

Photo 6.6   Building Damage in Beirut's Southern Suburbs

a. Destroyed buildings

b. Partially blasted buildings

c. Damaged buildings

Source: M. El Fadel and the study team.

El-Hermel areas. The volume of demolition waste transported as of that date was 5.75 million cubic meters, of which 1.43 million cubic meters were in Beirut, 3.32 million cubic meters were in the South, and 1 million cubic meters were in Bekaa (PCM 2007). This indicates that previous calculations were extremely conservative and underestimated the actual volumes being removed. Therefore, this chapter uses the numbers reported by the PCM to calculate the environmental damage cost due to hostilities.

Table 6.12   Estimated Volume of Demolition Waste

Waste type and location	Volume	Assumptions
Surface area per apartment	150–200 m²	
Total walls (inner, outer, pillars)[a]	45–90 m³	10–15% of surface area 3m as average height
Unit slab[a]	45–60 m³	0.3 m average thickness
Furniture and personal belongings[a]	18.75–50 m³	25–50% occupancy of surface area 0.5 m as average height
Total per housing unit	**109–200 m³**	
In Beirut's southern suburbs		
Number of units destroyed[b]	9,000	300 buildings, 10 stories, 3 apartments/story
Number of units partially blasted[c]	840	30% of the housing unit damaged
Number of units damaged[c]	2,100	5% of the housing unit damaged
Area's total demolition waste	**1,020,000– 1,871,000 m³**	
In the South		
Number of units destroyed[d]	8,791	
Area's total demolition waste	**956,000– 1,758,000 m³**	
In Baalbek El-Hermel		
Number of units destroyed[e]	375	
Number of units badly damaged[e]	400	30% of housing units damaged
Area's total demolition waste	**54,000– 99,000 m³**	
TOTAL FROM ALL AREAS	**2,030,000– 3,728,000 m³**	

Source: Authors' estimates in addition to data sources noted below.
Note: Data are based on field visits conducted during the preparation of this report (October and November 2006). m² = square meters, m³ = cubic meters.
a. Expert opinion, field visits.
b. GoL 2006b.
c. Dr. M. Fawaz, Department of Architecture and Graphic Design, American University of Beirut, pers. communication, November 2006.
d. Municipalities of South Lebanon.
e. M. El Jammal, head of Baalbek Municipality, pers. communication, October 18, 2006.

Waste loading, transport, and disposal. Although demolition waste is usually landfilled, the corresponding landfills are generally not subject to the same regulatory procedures as municipal solid waste (MSW) landfills because the latter contain mostly inert material.[57] While the cost of land-filling the demolition waste is relatively lower than that of MSW, some components of the demolition materials may be recycled, such as concrete, asphalt, metals, and wood.

In Lebanon, immediate removal and disposal of demolition waste was required to allow for reconstruction activities. Concerned municipalities, together with the Council for Development and Reconstruction in Beirut, the Ministry of Public Works and Transport in Baalbek, and the Council of the South in the Southern Districts, identified disposal sites for each region and contracted the excavation, hauling, transport, and disposal of the demolition waste.

In Beirut's southern suburbs, the collected demolition waste has been haphazardly dumped at four sites—two in low-lying areas near the sea, one off the road within the Choueifat cadastral area, and a temporary dump site along the Airport Road within the Bourj Al Barajneh cadastral area, as seen in photo 6.7.

The slope of the deposited waste has reached almost a 1-to-1 ratio, which could pose a safety hazard in the absence of adequate stability-control measures. Sea encroachment occurred to a minimum extent on the dump sites by the sea. Wherever this encroachment occurred, the bulky nature of the demolition waste gave it a relatively good angle of stability, minimizing the likelihood of its collapse into the sea. Although this invariably damaged the coastal ecosystem, the impact is difficult to quantify monetarily.

In the South, some municipalities where little demolition waste was generated used the waste to fill depressions in the roads or to reconstruct other building sites. In towns where large volumes of demolition waste were

Photo 6.7   Rubble Disposal Site in Beirut

Source: M. Sarraf.

generated, it was disposed of on nearby lands. Such is the case of Al Khyam, where the municipality threw the rubble in a nearby valley, 300 meters off the main road, facing the Israeli border. In Bint Jbeil, the municipality disposed of part of the waste in a valley 2 kilometers off the main road. However, the neighboring municipality of Aytaroun intervened and requested that the waste be directed to its own town to fill a seemingly abandoned pond, as shown in photo 6.8. The municipality of Maroun el Ras also sent its demolition waste to Aytaroun.

The damage associated with dumping of demolition waste in valleys and ponds goes beyond the cost of land, to include ecosystem damage and visual intrusion. However, the latter costs are difficult to quantify monetarily. Impacts on hydrology and hydrogeology are equally difficult to quantify monetarily but are expected to be limited because a considerable portion of the household hazardous material was removed prior to transport to the dump sites (UNDP 2007). Similarly, the opportunity cost associated with land use at dump sites is difficult to quantify but is not expected to be high, particularly in Khyam because of its proximity to the Israeli border.

In Baalbek El-Hermel, the collected waste was dumped in an abandoned quarry and in several other locations in the suburbs of Baalbek. Some waste was also used to rehabilitate land depressions caused by the military aggression.

Field visits revealed that sorting of some material such as construction steel, asbestos mats, and concrete bricks took place at the dump sites,

Photo 6.8   Aytaroun Pond Filled with Demolition Waste

Source: M. El Fadel and the study team.

primarily to recover steel for recycling. Other waste components such as personal belongings, furniture, and white goods could not be sorted because of the intensity of the destruction.

Asbestos mats were encountered mainly in Baalbek El-Hermel, as seen in photo 6.9. No asbestos was found upon preliminary site assessment as part of the UNDP initiative in the Beirut's southern suburbs and the South. No official reports on asbestos contamination of the waste have been published to date. Moreover, visual site inspections for asbestos-containing materials suggest that asbestos contamination is not a major issue of concern (UNDP 2007). As such, while friable asbestos poses a potential occupational hazard, the limited short-term exposure during the postconflict period is not likely to cause the development of severe health implications among the workers. Again, it is difficult to assign a monetary value to this impact, especially following such a short-term exposure.

Based on field visits and interviews, the equivalent unit cost of hauling and transporting the demolition waste is estimated at US$2.38 per cubic meter, as shown with the other cost estimates in table 6.13. Accordingly, the total cost of hauling and transport of the generated demolition waste is around *US$13.7 million.*[58]

Road depreciation. Two sets of roads can be identified in terms of damage from the movement of trucks transporting demolition waste: (a) roads in the South and in Baalbek El-Hermel and (b) roads in the Beirut area. The transport of demolition waste did not significantly affect the roads in the South and in Baalbek El-Hermel. Thus the damage to road infrastructure in these areas can be directly attributed to the military aggression.

Photo 6.9   Demolition Waste in Baalbeck El-Hermel

a. Asbestos mats | b. Steel waste

Source: M. El Fadel and the study team.

Table 6.13   Estimated Cost of Loading and Transporting Demolition Waste

Cost type	Estimate
Waste hauling[a]	
Bulldozer charging rate[b] (US$/day)	400
Filling capacity of 3 bulldozers[c] (truck/day)	30
Capacity per truck (m^3)	18
Daily volume of demolition waste loaded (m^3/day)	540
Cost of loading each truck (US$/m^3 of demolition waste)	**0.07**
Waste transport[a]	
Truck charging rate[b] (US$/day)	250
Daily number of round trips	6
Loading capacity per truck (m^3)	18
Daily volume of demolition waste transported per truck (m^3/day)	108
Cost of transport (US$/m^3 of demolition waste)	**2.31**
Total unit cost (US$/m^3)	**2.38**
Cost in Beirut's southern suburbs (US$ millions)	**3.4**
Cost in the South (US$ millions)	**7.9**
Cost in Baalbek El-Hermel (US$ millions)	**2.4**
TOTAL COST (US$ millions)	**13.7**

Source: Authors' calculations.
Notes: All cost estimates are from the year 2006. m^3 = cubic meters.
a. Based on field surveys and expert opinion.
b. Including wage of driver.
c. Average number of bulldozers per site, based on field surveys and expert opinion.

In contrast, the road infrastructure in the Beirut area was negatively affected by the high number of trucks required to move the significant demolition waste volume concentrated in Beirut's southern suburbs. Based on Geographical Information Systems (GIS) analysis and field surveys, table 6.14 shows that the estimated road refurbishment cost in Beirut due to the damage of demolition waste transport ranges between *US$240,000 and US$720,000.* These estimates are associated with a high level of uncertainty because it is difficult to separate the damage related directly to the hostilities from that related to truck travel in Beirut's southern suburbs.

Traffic delays. In Baalbek El-Hermel and in the South, traffic delays due to transportation of demolition waste were not encountered because of traffic rerouting away from the city center and because the dump site was on the city outskirts. In contrast, in Beirut's southern suburbs, the increased number of trucks on congested roads caused many people using the southern corridor of Beirut to spend an extra one to three hours in traffic.

Table 6.14   Cost of Road Depreciation in Beirut

Cost factor	Estimate
Average road length (km)[a]	2–3
Average road width (m)[a]	6–8
Average road area (m²)	12,000–24,000
Cost of road refurbishment (US$/m²)[a]	20–30[b]
TOTAL COST (US$)	**240,000–720,000**

Source: Authors' calculations.
Note: Data are based on 2006 estimates. km = kilometer, m = meter, m² = square meter.
a. Based on field surveys, expert opinion, and GIS analysis.
b. Range accounts for degree of intervention and thickness damaged.
Road materials are 40 centimeters of compacted gravel and 10 centimeters of asphalt.

While it is difficult to differentiate between delays related to trucks transporting demolition waste and delays related to bombed roads in Beirut, the delays arguably would have been much shorter without truck movements. Therefore, a delay of two hours per day is attributed to the transport of demolition waste.

This delay translates to an economic loss of about *US$51 million to US$68 million*, in terms of wages and fuel, as detailed in table 6.15. A factor of 0.5 is applied to account for opportunity cost versus actual cost, assuming that half of the time lost in traffic delays is productive. The other half could be translated into an impact on quality of life, which was not estimated in monetary terms. In an assessment of these estimates, the following factors are also worth noting:

- The road structure (narrow and poorly maintained roads) exacerbates traffic congestion.
- Trucks carrying debris were working around the clock with practically no alternate routes or sites, particularly in Beirut's southern suburbs.
- Working off-peak was not an option, even when the hauling schedule decreased to 12 hours or less.
- A proper public bus transit system does not exist in the area, and general traffic is dominated by passenger trips or shared taxis.

Cost of land for waste disposal. We assume conservatively that all the waste in each area is disposed of in one equivalent landfill with a height of 25 meters and a buffer zone of 30 percent. By the estimates shown in table 6.16, the cost of land needed for disposing of the demolition waste is around US$78,000 in Baalbek El-Hermel, US$1.7 million in the South, and US$74 million in Beirut—totaling *US$76.9 million.* These numbers are

Table 6.15   Estimated Cost of Traffic Delays

Type of delay or cost	Estimate
Average extra time spent in traffic (hours/day)[a]	2
Average hourly wage (US$/hour)[b]	2.5
Number of working days/month[c]	22
Fraction of lost productive time[a]	0.5
Duration of waste removal (months)	6–8
Opportunity cost of time (US$/person/6–8 months)	330–440
Average daily number of affected commuters[d]	115,150
Opportunity cost of time (US$ millions)	**38–51**
Fuel consumption per hour in traffic (liter/hour)[a]	1.0
Unit cost of fuel (US$/liter)[a]	0.8
Cost of fuel spent (US$/person/month)	35.2
Number of affected vehicles[d]	60,000
Cost of fuel spent (US$/person/6–8 months)	211–282
Fuel cost spent per 6–8 months (US$ millions)	**13–17**
Total cost of traffic delays (US$ millions)	**51–68**

Source: Authors' calculations.
Note: Costs are based on 2006 estimates.
a. Based on field surveys and expert opinion.
b. Based on GDP of US$5,300 per capita.
c. Peak travel delays are assumed to occur 22 working days per month.
d. DMJM+HARRIS 2003.

Table 6.16   Estimated Cost of Land for Demolition Waste Disposal

Location	Waste volume (m³ thousands)	Landfill height (m)	Waste area (m² thousands)	Landfill area (m² thousands)	Land cost[a] (US$/m²)	Total land cost (US$ millions)
Beirut's southern suburbs	1,430	25	57.2	74.4	1,000	74.4
South	3,320	25	132.8	172.6	10	1.7
Baalbek El-Hermel	1,000	25	40.0	52.0	15	0.8

Source: Authors' calculations.
Notes: Costs based on 2006 estimates. m = meter, m² = square meters, m³ = cubic meters.
a. Estimated land cost per square meter is based on real estate information and expert opinion.

underestimates because, in reality, there are more sites and the height of the waste is often lower than 25 meters, resulting in a need for larger areas.

Because most selected sites are not in prime locations, the unit cost of land adopted in the estimations is average to low. In the case of Beirut, for example, the unit cost of land by the sea ranges between US$2,000 and

US$5,000 per square meter. Yet, the selected waste sites are by the airport and hence of lower value.

Disposing of demolition waste on land valued at around US$1,000 per square meter may seem an enormous cost. However, it is important to note that in Beirut, alternative dump sites near the damaged sites were extremely difficult to locate, if not nonexistent. Hauling rubble from Beirut to cheaper sites in the Bekaa, the South, or the North of the country was not feasible at the time, given that the infrastructure and most connecting bridges had been destroyed. In addition, political pressure to keep the waste within the area of the southern suburbs of Beirut was mounting because of the potential benefit from steel recycling. Although the income from the latter should theoretically be deducted from the overall damage estimates, this was not possible for lack of well-documented information about recycling activities.

In the South, the pond and the valley are considered as prime lands, and road depressions have minimal direct cost. In Baalbek El-Hermel, lands along the roads are of considerable value, and quarries are of lower value.

Depreciation of land surrounding dump sites. Disposal of rubble and debris in various dump sites, especially in those along the coast, represents a health hazard to the surrounding neighborhoods as well as a visual intrusion that affects the quality of life. Within the context of this analysis, it was not possible to assess the effect of the dump sites on the value of surrounding land.

Overall, the estimated damage caused by the presence of large quantities of rubble and debris ranges within *US$142 million to US$159 million*, as table 6.17 summarizes.

Table 6.17   Estimated Damage Costs of Demolition Waste
US$ millions

	Damage cost, by location			
Type of demolition waste	*Beirut's southern suburbs*	*South*	*Baalbek El-Hermel*	*Total*
Waste loading, hauling, and transport	3.4	7.9	2.4	13.7
Road maintenance	0.2–0.7	0.0	0.0	0.2–0.7
Traffic delays	51.0–68.0	0.0	0.0	51.0–68.0
Land for disposal	74.4	1.7	0.8	76.9
Land depreciation	n.a.	n.a.	n.a.	n.a.
Total cost	**129.0–146.5**	**9.6**	**3.2**	**142.0–159.0**

Source: Authors' calculations.
Note: Costs are based on 2006 estimates. m^3 = cubic meters, m^2 = square meters, n.a. = not available.

Damage Costs due to Military Waste

The impacts of military waste are estimated in terms of human deaths and injuries, limited access to agricultural land, and demining efforts. Given the complexity of the demining process and its associated delays, the economic valuation uses a period of two years as a minimum time lag for demining completion.

Deaths and injuries. According to the Lebanese National Demining Office, from August 14, 2006, to April 03, 2007, UXOs caused 224 casualties, including 29 deaths and 195 injuries (MACCSL 2007). The number of UXO casualties by the end of the two-year study period (August 2006 to August 2008) is projected based on the casualty trends as of April 2007. The distribution of the projected numbers in terms of mortality, morbidity, and age is assumed to be similar to the current distribution. Overall, the UXOs caused an estimated 36 deaths and 239 injuries.[60]

We estimate the damage cost of premature mortality and morbidity from UXOs based on DALYs. Illnesses are weighted by severity such that a relatively mild illness or disability represents a small fraction of a DALY, while a severe illness represents a larger fraction of a DALY. One year lost to premature mortality represents 1 DALY, and future years lost are discounted. For injuries resulting from UXOs, such as leg or arm amputation, the disability weight adopted to assess damage cost is 0.3 (Murray and Lopez 1996).

Using the HCA approach, the lower value of 1 DALY is equal to the GDP per capita, or US$5,300. The upper value of 1 DALY is estimated at US$42,000, based on the VSL divided by a time horizon of 25 years and a discount rate of 4 percent.[61] Table 6.18 presents the calculation of the damage cost due to mortality and morbidity by age group. Accordingly, the estimated damage cost of casualties resulting from UXOs ranges between *US$14 million and US$109 million.*

Access to agricultural land. The limited access to agricultural lands in the South imposed by UXOs was expected to have a significant impact on agricultural production and farmer livelihoods for at least two years. Farmers may respond to this crisis in various ways. They may burn their orchards to eliminate the UXOs, losing their plantations in the process; they may simply wait for their lands to be cleared of the UXOs; or they may migrate to urban areas. Again, it is difficult to assign a monetary value on these types of behaviors. As such, the impact of the UXOs on farmers is estimated by assessing only the loss in production due to lack of access to agricultural lands.

Table 6.18   Estimated Damage Cost of Unexploded Ordnances (UXOs)

Age group (years)	Current number of casualties[a]	Current and projected number of casualties[b]	DALYs per case[c]	Value of 1 DALY (US$ thousands)	Current and projected economic loss (US$ millions)
Mortality					
0–12	2	2.5	33	5.3–42.0	0.43–3.40
13–18	4	4.9	36	5.3–42.0	0.94–7.43
19+	23	28.2	20	5.3–42.0	2.99–23.72
Subtotal	**29**	**35.6**	**n.a.**	**5.3–42.0**	**4.36–34.55**
Morbidity					
0–12	24	29.5	9.9	5.3–42.0	1.55–12.25
13–18	39	47.9	10.8	5.3–42.0	2.74–21.72
19+	132	162.1	6.0	5.3–42.0	5.15–40.84
Subtotal	**195**	**239.4**	**n.a.**	**5.3–42.0**	**9.44–74.81**
TOTAL	**224**	**275.0**	**n.a.**	**5.3–42.0**	**13.80–109.35**

Source: Authors' calculations and sources of data as noted below.
Note: The monetary value of 1 DALY is based on the GDP per capita as the lower bound and the VSL as the upper bound. n.a. = not applicable.
a. MACCSL 2007.
b. The projected number of casualties is based on the same age distribution of current casualties.
c. Murray and Lopez 1996.

As figure 6.3 illustrates, however, cluster bombs were scattered throughout the South, making it difficult to quantify the total area of inaccessible agricultural lands.

The valuation is based on (a) estimating the total value of the annual production in the areas concerned and (b) calculating the share of this value lost to UXOs. In table 6.19, the annual production in the South and Nabatyieh is estimated at US$268 million.

Accordingly, we adopt two scenarios to estimate the damage cost to agricultural productivity. They assume that 10–25 percent of the South and Nabatyieh agricultural land[62] would be inaccessible during the first year, and 5–10 percent would be inaccessible during the second year. Based on these assumptions, table 6.20 estimates that the total loss in agricultural production due to UXOs ranges between *US$40 million and US$94 million* over a period of two years.

Demining. As of May 2007, the Lebanese Army, the UN Mine Coordination center, and some NGOs had worked hard to clear an estimated 14 million square meters affected by UXOs (PCM 2007). Cleaning of agricultural land started around February 2007, after giving priority to clearing schools, public roads, and housing areas. An annual budget of

Figure 6.3   Map of Cluster Bomb Strikes as of July 23, 2007

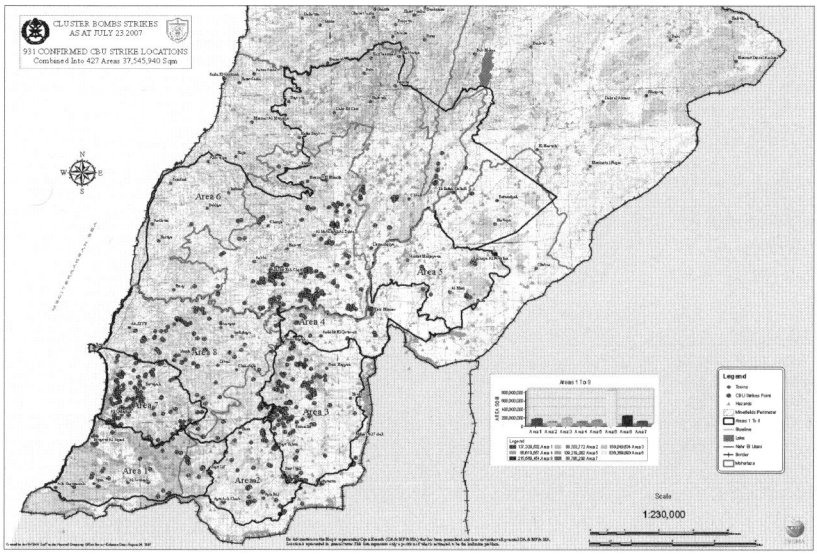

Source: Lebanon Mine Action Center.

Table 6.19   Estimated Annual Value of Agricultural Production in the South and Nabatiyeh

Crop type	Cultivated area (du thousands)[a]			Production rate[b] (ton/du)	Total production (tons thousands)	Value[c] (US$/ton)	Value (US$ millions)
	South	Nabatiyeh	Total				
Cereals	37.6	59.5	97.2	0.28	26.8	297	8.0
Legumes	2.1	5.8	8.0	0.54	4.3	565	2.4
Fruit trees	123.3	20.8	144.0	1.26	182.0	746	135.8
Olives	89.3	116.1	205.5	0.29	58.8	1,268	74.5
Oleaginous trees	5.8	3.8	9.6	0.10	0.9	2,083	1.9
Vegetables	20.8	12.1	32.9	3.19	104.9	251	26.3
Raw tobacco	14.6	40.0	54.7	0.12	6.4	2,988	19.1
TOTAL	**293.5**	**258.1**	**551.9**	**0.70**	**384.1**	**698**	**268.0**

Source: Authors' calculations and sources of data as noted below.
Note: 1 dunum (du) = 1,000 square meters.
a. Based on the 1999 agricultural census, Ministry of Agriculture. www.agriculture.gov.lb.
b. Based on the MOA 2004 production statistics and calculated by dividing the total production for each crop category by its cultivated area for the year 2004.
c. Average of the import and export values. The value per ton of import (export) is calculated by dividing the import (export) total value of each crop category by its quantity. Totals may not add up due to rounding.

Table 6.20   Estimated Range of UXO Damage Cost to Agricultural Productivity
US$ millions

	Assumptions for damage cost estimation under each scenario	Low-bound scenario	High-bound scenario
First year	Losses of 10% of annual production value under the low-bound scenario and of 25% under the high-bound scenario	26.8	67.0
Second year	Losses of 5% of annual production value under the low-bound scenario and of 10% under the high-bound scenario	13.4	26.8
TOTAL		**40.2**	**93.8**

Source: Authors' calculations.

US$5.5 million has been allocated to demining.[63] If one assumes two years of operations until demining would be complete, the total cost is estimated at *US$11 million*.

Overall, the damage caused by UXOs is estimated to range between *US$65 million and US$214 million*, as table 6.21 presents. It is a conservative estimate, because it does not consider other impacts such as that of UXOs on forests.

Damage Costs of Medical Waste

The increase in casualties and hospital bed occupancy generated an estimated 200 to 250 tons of medical waste (UNDP 2007). Health-care waste can be grouped into two broad categories: nonrisk waste and risk waste.[64] Because of the lack of data regarding the composition of the medical waste produced during hostilities, we assume that no segregation is practiced, and therefore the generated waste is infectious. This requires sterilization at a cost of US$60 per ton, followed by disposal with regular solid waste at an operational landfill in the country, at a cost of US$15 to US$120 per ton.[65] Therefore, the cost of handling the medical waste generated during the hostilities (including treatment, transport, and disposal) ranges between *US$15,000 and US$45,000*.

Shipping of hazardous waste in accordance with the Basel Convention on the Transfrontier Shipment of Hazardous Waste would cost US$10,000 per ton. However, this option has not been considered in this analysis, because, to the best of our knowledge, shipment of infectious medical waste has never been practiced in Lebanon.[66] Given the large quantities of

Table 6.21   Estimated Damage Cost due to UXOs
US$ millions

Damage type	Minimum	Maximum
Casualties	14	109
Loss in agricultural opportunities	40	94
Demining	11	11
TOTAL	**65**	**214**

Source: Authors' calculations.

medical waste generated by the hostilities, if the option of shipping and treating medical waste was considered by Lebanon, the overall damage cost resulting from medical waste would be substantially increased.

In addition, Lebanon received about 502 tons of medicines and medical aid supplies during and after the hostilities (PCM 2007).[67] Some pharmaceuticals may have arrived in Lebanon past their expiration dates, while others arrived in unwanted quantities (UNDP 2007). Safe disposal of unwanted or expired drugs often creates a major problem. However, because at the time of writing the quantity of unwanted and expired drugs was not known, it was not possible to account for their disposal cost. Due to the limitation mentioned above, the total estimated impact of *US$15,000 to US$45,000* likely underestimates the full impact of medical waste on the environment.

Summary: Damages due to Demolition, Military, and Medical Waste

The assessment of the impacts of demolition, military, and medical waste resulting from military aggression during the July–August 2006 hostilities are conservatively estimated to range from US$207 million to US$373 million. Overall, the impact averages to *US$290 million, or 1.4 percent of GDP in 2006* (World Bank 2007).

More than half of the estimated impact is due to demolition waste, particularly as a result of the high cost of land for waste disposal and the traffic delays in and around Beirut's southern suburbs. The costs of waste loading, transport and disposal, and demining are considerably lower, as presented in table 6.22 and figure 6.4.

The overall estimate is likely to undervalue the real impact caused by the hostilities on the waste sector because several potential effects have not been quantified. For example, the chapter does not account for the depreciation value of the real estate surrounding huge dump

Table 6.22   Estimated Cost of Damage due to Military Waste
US$ millions

	Damage cost			
Damage type	Minimum	Maximum	Mean	Mean (% of total)
Demolition waste	**141.8**	**159.3**	**150.5**	**52**
Waste loading, transport, and disposal	13.7	13.7	13.7	5
Road depreciation	0.2	0.7	0.5	0
Traffic delays	51.0	68.0	59.5	21
Land for waste disposal	76.9	76.9	76.9	27
Military waste (from UXOs)	**65.0**	**214.2**	**139.6**	**48**
Deaths and injuries	13.8	109.4	61.6	0
Access to agricultural land	40.2	93.8	67.0	0
Demining	11.0	11.0	11.0	4
Medical waste	**0.0**	**0.0**	**0.0**	**0**
Total	**206.8**	**373.5**	**290.2**	**100**

Source: Authors' calculations.

Figure 6.4   Impacts of the July–August 2006 Hostilities on Waste

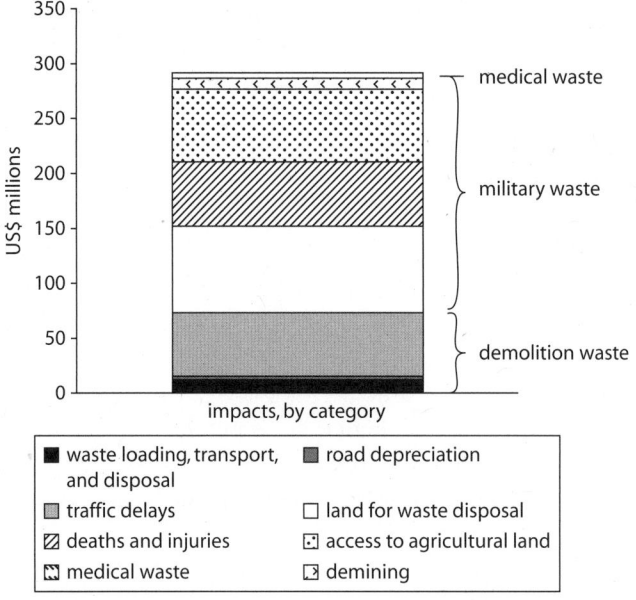

Source: Authors' calculations.

sites, the health hazard generated by the manipulation of the demolition waste, and the impact of UXOs on forests.

Conclusions

The previous sections estimated the damages associated with the oil spill and waste caused by the 2006 hostilities in Lebanon. In addition to these damages, the conflict affected other environmental categories and economic activities, such as water, forests, air quality, and quarrying. The World Bank (2007) estimated the total cost of environmental degradation (COED), which includes the costs of all the hostility-related impacts of environmental degradation (except for those related to air quality, due to insufficient information). Table 6.23 indicates that the total COED caused by the 2006 hostilities in Lebanon is between US$527 million and US$931 million, averaging at *US$729 million, or 3.6 percent of GDP* in 2006.

It is interesting to note that a previous study estimated the annual COED in 2000 at US$565 million, or 3.4 percent of GDP (Sarraf, Larsen, and Owaygen 2004).[68] A comparison between the two estimates shows that the environmental damage caused by the 34-day hostilities was higher than that caused in a whole year in peacetime. On average, the environmental damage caused by one day of hostilities costs about US$21.5 million, compared with US$1.7 million in damage costs per day in peacetime.[69]

Figure 6.5 illustrates the main components of the damage cost attributable to the 2006 hostilities. The largest estimated impact is on the *waste*

Table 6.23   Estimated COED of the 2006 Hostilities in Lebanon

Impact category	Minimum (US$ millions)	Maximum (US$ millions)	Mean (US$ millions)	Mean (% of GDP)[a]
Waste	206.8	373.5	290.2	1.4
Oil spill	166.3	239.9	203.1	1.0
Water	131.4	131.4	131.4	0.6
Quarries	15.4	175.5	95.5	0.5
Forests	7.0	10.8	8.9	0.0
Air[b]	—	—	—	—
Total COED caused by the hostilities	**526.9**	**931.1**	**729.0**	**3.6**

Source: Authors' calculations; World Bank 2007.
Note: — = not available.
a. Based on an estimated GDP for 2006 of US$20.5 billion (Economic Intelligence Unit 2006).
b. The impact of hostilities on air quality could not be estimated because of insufficient data.

Figure 6.5   Estimated COED in Lebanon, by Category, 2006

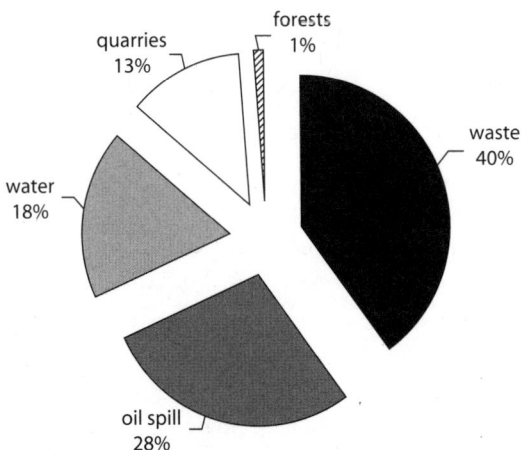

Source: Authors' calculations.
Note: Percentages represent the estimated share of the US$729 million COED caused by the Lebanon conflict and attributable to each damage category.

sector, mainly due to the high costs of disposing of the waste attributable to demolition, casualties, and agricultural losses caused by UXOs in South Lebanon. The impact of the *oil spill* is the second most important, primarily because of the high costs of cleaning the oiled waste, the cost of burned and spilled oil, and the income losses from coastal services (for example, beach resorts). The impacts of the hostilities on water, quarries, and forests are comparatively less important, although they are conservatively valued. Once again, the limited data availability makes these estimated costs only indicative of the real value of damages.

The damages that the hostilities caused to the waste sector also exacerbated long-lasting environmental issues, such as waste management concerns. Lebanon can take several measures to resolve this problem, including (a) agreeing to allocate land for sanitary landfills in different Mohafazats; (b) enacting the Integrated Waste Management Law and implementing the National Solid Waste Strategy in a competitive and cost-effective manner; and (c) providing incentives to municipalities to treat their waste.[70]

The damage caused by the 2006 oil spill revealed the need to develop and implement sustainable oil spill preparedness and response systems. A national oil spill control and contingency plan should be in place and exercised periodically with the various government agencies and all relevant stakeholders. It should also include a joint effort at capacity building by key stakeholders such as local industry, government, and NGOs to accommodate for local conditions.

Notes

Maria Sarraf is a Senior Environmental Economist at the World Bank, South Asia Region, MC10-1019 1818 H Street, NW, Washington, DC 20433, USA.

Lelia Croitoru is an Environmental Economist at the World Bank, Middle East and North Africa Region, 1818 H Street, NW, Washington, DC 20433, USA.

Mutasem El Fadel is Professor and Dar Al-Handasah (Shair & Partners) Chair in the Engineering Department of Civil and Environmental Engineering, American University of Beirut, P.O. Box 11-0236 / FEA-CEE, Riad El Solh / Beirut 1107 2020, Lebanon.

Karim El-Jisr is Director, ECODIT Liban SARL, Rue Hôpital Sacré-Coeur, Centre Ghaleb, 1er Etage, Baabda, Liban.

Erkki Ikäheimo is a Senior Consultant, Kalakontintie 1 B 38, 02230 Espoo, Finland.

Erich Gundlach is President, E-Tech International Inc., 15 River Park Drive, New Paltz, NY 12561, USA.

Samia Al-Duaij is Operations Officer at the World Bank, Middle East and North Africa Region, 1818 H Street, NW, Washington, DC 20433, USA.

1. The estimate was based on several assessments, including the one by Khatib and Alami (2006).

2. Communication with the Lebanon Ministry of Environment (MOE), August 2007.

3. According to tests done by CEDRE, the oil spilled appeared to be an intermediate fuel oil (IFO) with a viscosity of 150 CentiStokes (cSt) at 50°C. www.cedre.fr.

4. The MEDSLIK computer model can calculate the oil spill trajectory and fate developed for the eastern Mediterranean. The European Union helped to obtain the MEDSLIK model from the Oceanography Centre at the University of Cyprus.

5. These indirect effects are much harder to determine. While an oil-covered bird is clearly affected, partial oiling gives the appearance that the bird "survived" the spill, only to die sometime later from oil ingested during preening or from other complications. In addition, because a spill also damages the birds' food sources (shoreline organisms and fish), the affected areas cannot sustain the previous level of bird population. This injury may occur over a period of several years after the spill and, therefore, is particularly difficult to assess.

6. Interview in April 2007 with Dr. Ghassan Ramadan-Jaradi, professor of ornithology and chair of the Palm Islands Nature Reserve GAC.

7. Observed during a field visit in October 2006.

8. UNEP did not discover high concentrations of soluble hydrocarbons. However, dense nonaqueous phase liquids (DNAPLs) and the high volatility of hydrocarbons may have caused the contamination to go unobserved.

9. For example, 10 years after the 1992 Aegean Sea spill that released 70,000 tons of oil into a harbor of Galicia, Spain, the area had still not fully recovered (Chas-Amil and others 2004). It has also been suggested that recovery of different species of shellfish may take anywhere from a few months to more than 12 years. In 2002, the damage provoked by the Prestige spill of 64,000 tons of oil off the Galician shore had a severe impact on local biodiversity, and recovery was predicted to take between 2 and 10 years (Loureiro and others 2006).

10. For example, the 1989 Exxon Valdez oil spill in Alaska provoked a sudden death of fish three years after the accident (Fall and others 2001). The Aegean Sea spill caused a 33 percent fall in catch no less than six years after the event (Chas-Amil and others 2004).

11. Different sources suggest different impact time frames, such as 10 to 50 years' serious impact on marine biodiversity and 1 to 10 years' catastrophic impact on the littoral zone (UNDP 2007), or a few months to several years' impact on tourism (director general Ministry of Tourism, pers. communication, 2007).

12. This appeared credible in April 2007, when it was observed that little or no oil existed in the water column, and a good portion of sandy beaches had been cleaned.

13. This is a conservative time frame because it does not capture potential effects either not yet probed or yet to occur over an extended period of time.

14. A review of oil spill studies of the Exxon Valdez (Cohen 1995; Monson and others 2000; Carson and others 2003); the Prestige (Loureiro and others 2006); the Erika (Bonnieux and Rainelli 2002); and the Amoco Cadiz (Grigalunas and others 1986) indicates that none of them was related to conflicts. Thus, available information does not give any example or guidance on how to separate the extent of damage according to the cause because no combined causes are involved. An exception is the 1991 Gulf war, which triggered an oil spill in Kuwait, for which the economic analysis of the environmental damages was limited to remedial approaches.

15. Interview at the Syndicate of Hotel Owners, September 2006.

16. Room rates vary from US$40 to US$300 per night, averaging US$100 per night. Meals, phone, and laundry average US$50 per day.

17. Based on internal statistics of the Ministry of Tourism.

18. According to April 2007 interviews, the daily price per apartment varies widely, depending on the quality of services, number of bedrooms, the season, and other factors. For example, during the low season, a one-bedroom apartment costs between US$55 per night (in Savoy Suites, Raouche Beirut, April 2007) and US$175 per night (Lahoya Homes, Manara, Beirut, April 2007), while a two-bedroom apartment costs between US$200 and US$450 per night (Lahoya Homes, Manara, Beirut).

19. A typical example is the Bamboo Bay resort in Jiyeh.

20. Based on the Lebanese Ministry of Tourism publication, *Yellow Pages Tourism 2006*.

21. Visitor spending ranges between US$7 and US$30 per day (interview with Syndicate of Beach Resorts representative, October 2006).

22. A typical example is the Rimal complex resort in Jounieh.

23. Based on the Lebanese Ministry of Tourism publication, *Yellow Pages Tourism 2006*.

24. Based on interviews at a random sample of 10 chalet complexes outside Beirut, April 2007.

25. The average is based on the number of day-visits: 1,500 on working days, 5,000 on Saturdays, and 20,000 on Sundays (Cedars for Care, pers. communication, April 2007).

26. This value does not reflect what individuals actually pay to visit public beaches, because they are usually free of charge. Rather, the benefit amount reflects the economic value in terms of willingness to pay (WTP) to enjoy the beach. Because no information on the WTP for public beaches is available, we use data on the WTP for private beaches (US$20). The services provided by private beaches are of better quality than those of public beaches; therefore, the WTP for public beaches is assumed to be around 50 percent of that of private beaches.

27. 8,400 visitors per day × 30 days × US$10.

28. The cost of boat cleanup is captured in the total cleanup costs at the end of this section.

29. The overall losses due to the oiled fishing boats are reflected through the loss in fish catch and the maintenance costs paid for one year (2006). Because the loss in fish catch is already accounted for in the "Fishing" subsection, this section considers only the maintenance costs to avoid double-counting.

30. The interviews suggested that there are about 20 to 30 boats per small marina (for example, Miramar and Las Salinas), 80 to 100 boats per medium-size marina (for example, Movenpick and Halat sur Mer), and about 300 to 400 boats per large marina (for example, Dbayeh and ATCL).

31. According to interviews at the Power Marine and Dolphin Team boat suppliers in Mount Lebanon, boat prices vary from US$10,000 to US$500,000, depending on size and model. The lifetime of a boat is usually 20 to 30 years. Because boats often become obsolete long before the end of their lifetimes, it is reasonable to assume a lifetime of 20 years.

32. Interviews at five marinas show that the cost per meter per season varies widely: for example, US$400–US$700 per meter per season in Marina Dbayeh, US$300–US$400 in Holiday Beach, and US$2,500–US$3,000 in Movenpick.

33. Interview with Dr. Ghassan Ramadan-Jaradi, chair of the Palm Islands Nature Reserve GAC, April 2007.

34. Interview with Dr. Ghassan Ramadan-Jaradi, chair of the Palm Islands Nature Reserve GAC, April 2007.

35. We found no studies estimating the damage cost of dead or injured turtles in Lebanon as a result of the oil spill. Whitehead (1993) found a WTP of about US$11 per year to ensure the continued existence of the loggerhead sea turtle (*Caretta caretta*) in the United States. In another study, Whitehead (1992) estimated the WTP of about US$44 per person for continued existence of turtles for the next 25 years. It is difficult, however, to apply these estimates in Lebanon's case because (a) only a small (unknown) fraction of the total WTP can be attributed to the loss of the three loggerheads, and (b) these estimates do not reflect the country's specific biodiversity conditions.

36. Alternatively, benefits transfer could be used. Brown (1992) estimated a range of values from US$167 per gull to US$6,000 per peregrine falcon. Loureiro and others (2006) estimated a cost of about US$250 per dead bird. That would correspond to about US$60 per dead bird in Lebanon, after adjustment to GDP ratio. Assuming that all oiled birds found in Lebanon would die, as in many other oil spill incidents (McCay and others 2004), the total value of the 92 injured birds would be about US$5,500. However, because of the method's inability to reflect the specific biodiversity conditions in Lebanon, this estimate will not be considered in the analysis.

37. It is known that there are 3.5 white pelicans per square kilometer and 3.8 squacco herons per square kilometer. There was one injured white pelican and four injured squacco herons. (Ramadan-Jaradi and Ramadan-Jaradi 2001; MOE 2004)

38. The study, which aims to develop indicators for future monitoring, is funded by the Italian government through the World Conservation Union (IUCN) and implemented by the American University of Beirut (communication with Ms. H. Kilani of IUCN, Lebanon, April 2007).

39. The monitoring program includes the following activities: developing measurable indicators, setting guidelines for monitoring, analyzing data on a yearly

basis, and developing a set of guidelines for use in future incidents (interview with Dr. Ramadan-Jaradi, April 2007, and communication with Ms. H. Kilani of IUCN, April 2007).

40. This figure does not cover either the cleanup cost already spent of US$85,600 at the time of this writing in April 2007 (Dr. Jaradi communication; SDC 2007), or the estimated cost of the second phase of cleanup and monitoring operations, estimated at about US$1 million (according to MOE). Both of these costs are accounted for in the "oil spill cleanup" subsection.

41. The Byblos tour operators are Tania Travel, Kurban Tours, Nakhal, Anastasia, Ariane, Rida Travel, and Wild Discovery.

42. Visitor cost estimates are based on interviews with tour operators Tania Travel, Kurban Tours, and Nakhal.

43. This is the minimum spending on souvenirs and boat rental, based on the judgment of local experts.

44. Navrud and Ready (2002) make a comprehensive review of studies valuing the WTP to conserve sites with cultural heritage, such as monuments and archaeological sites. According to that review, the annual WTP ranges from as little as US$0.60 to US$1.00 per household to preserve Bulgarian monasteries to as much as US$134.00 per household to conserve the recreational value of aboriginal rock paintings in Nopimi Park, Canada. These estimates represent the WTP to protect historical sites from pollution other than oil.

45. Interviews, October 2006 and April 2007.

46. Communication with Dr. M. Nader at the Marine Resources and Coastal Zone Management Program, Institute of the Environment, University of Balamand, October 2006.

47. Interview at the Syndicate of Fishermen, October 2006.

48. Interview at the University of Balamand, October 2006.

49. Estimate based on data provided by the MOE in August 2007.

50. See World Bank (2007), table 2.24, for a detailed list of organizations, activities, and estimated cleanup costs.

51. The second phase of oil spill cleanup is assumed to generate an additional 4,500 cubic meters of solid waste, based on surveys of the remaining polluted sites conducted along the Lebanese coast (MOE 2007c). Because there is no information about the management of oil waste that would be generated under the second phase, the cost of managing the 4,500 cubic meters of solid waste is not accounted for in the study.

52. UNDP (2007) indicates a cost of US$25,000 for treating 2,400 cubic meters of low- to medium-contaminated sand and pebbles.

53. UNDP (2007) estimates that shipping 100 cubic meters under the Basel Convention costs about US$1 million.

54. Information from E-MINE (Electronic Mine Information Network). "Mine Action Co-ordination Centre South Lebanon: Unexploded Ordnance Fact Sheet." http://www.mineaction.org/overview.asp?o=540.

55. The analysis does not account for the potential treatment of demolition waste because it did not enter the actual handling scenario. For information about options for treatment and disposal of the demolition waste, one can refer to the extensive analysis carried out by UNDP (2007).

56. Information on the WHO Web site further defines disability-adjusted life years: http://www.who.int/healthinfo/global_burden_disease/metrics_daly/en/.

57. Demolition waste results from military activities rather than from a systematic, well-organized effort that produces mostly inert demolition material. Therefore, it can be easily associated with potential subsurface pollution due to rainwater infiltration.

58. The average hauling distance was about 2 to 3 kilometers, whether in suburbs of Beirut, the South, or the Bekaa. Therefore, contracts with hauling contractors implicitly took the distance into consideration and appeared to have been based mainly on the volume hauled, set as a function of truck capacity. This may explain the difference between the estimated volumes and the reportedly hauled volumes, taking into consideration the density variations.

59. "Solidere" stands for *Société libanaise pour le développement et la reconstruction de Beyrouth* (The Lebanese Company for the Development and Reconstruction of Beirut). http://www.solidere.com/solidere.html.

60. Projections for the two-year period are made because of the early writing of the original report (April 2007), which did not allow for information gathering on the ground for future years. As of February 2010, the United Nations Interim Force in Lebanon reports that cluster munitions and UXOs killed 27 and injured 234 civilians in South Lebanon, and accidents during demining caused another 14 deaths and 41 injuries of demining personnel. Since spring 2008, the civilian casualties dropped significantly in the range of 0 to 2 per month (http://unifil.unmissions.org/Default.aspx?tabid=1519). Accordingly, cluster munitions and UXOs had caused 41 deaths and 275 injuries until spring 2008. These numbers are in line with the projections made in April 2007 (36 deaths and 240 injuries, as reported in table 6.18). Because no information is available concerning the distribution by age of the updated data, this chapter uses the information collected in April 2007 for the estimation.

61. This value is based on the meta-analysis by Viscusi and Aldy (2003), who suggested a VSL between US$5.5 million and US$7.6 million in 2000. This value is adjusted to Lebanon, taking into account the GDP per capita differential and the inflation rate.

62. Evenly distributed among crop categories in the two Mohafazas, South, and Nabatiyeh.

63. Interview with the Director of Lebanese Army National Demining Office, April 2007.

64. "Nonrisk" waste, as defined by WHO, is estimated to constitute 75 percent or more of the total hospital waste stream—in some cases, 90 percent. Nonrisk waste, which comprises general domestic waste components generated primarily from domestic and administrative services, is considered to represent no potential risk from infectious, chemical, or other properties associated with health care. If waste separation and segregation are undertaken properly, nonrisk waste can be disposed of similarly to municipal waste. The remaining 10–25 percent of hospital waste comprises those components that are potentially contaminated with material associated with infectious, chemical, or other hazardous characteristics. WHO defines waste with such characteristics as "risk waste" that must be handled and disposed of in a way that minimizes the potential for human exposure and contamination. WHO has classified risk waste into seven distinct environmental categories: infectious, pathological, sharps, pharmaceutical, chemical, pressurized containers, and radioactive wastes.

65. Cost estimate is based on landfilling charges in 2006 at Zahlé and Nehmé landfills.

66. The American University of Beirut ships selected hazardous waste (but not infectious medical waste) on a periodic basis.

67. Medical aid included anesthetics, antibiotics, anti-convulsants, anti-depressants, anti-diarrheals, anti-fungals, anti-inflammatories, anti-retrovirals, heart medicines (anti-arrhythmic, anticoagulant, anti-cholesterol, anti-platelet & hypertension treatments), laxatives, stomach medicines (beta-blockers and H2- antagonists), painkillers and tranquilizers.

68. The study estimated the COED due to air pollution at 1.1 percent of GDP; water degradation, at 1 percent; land and wildlife degradation, at 0.6 percent; coastal zone degradation, at 0.7 percent; and waste management, at 0.1 percent.

69. Adjusted from 2000 to 2006 using the GDP deflator.

70. Financial incentives such as carbon finance can alleviate the financial burden imposed by the waste sector.

References

BIL (BOTAŞ International Ltd). 2003. "Coastal Sensitivity Maps and Containment Manual, Iskenderun Gulf and Vicinity." Document BOT-REP-ENM-GEN-12 prepared for BTC Company's Baku-Tbilisi-Ceyhan (BTC) Crude Oil Pipeline Project, BIL, Ankara, Turkey.

Bonnieux, F., and P. Rainelli. 2002. "Évaluation des dommages des marées noires: une illustration à partir du cas de l'Erika et des pertes d'agrément des résidents." Economie et Statistique 357 (1):173–87.

Brown, G. 1992. *Replacement Costs of Birds and Mammals*. Seattle: University of Washington.

BTC (Baku Tbilisi Ceyhan). 2003. "Coastal Sensitivity Maps and Containment Manual, Iskenderun Gulf and Vicinity. BTC Crude Oil Pipeline Project, Ankara, Turkey.

Carson, R. T., R. C. Mitchell, M. Hanemann, R. J. Kopp, S. Presser, and P. Ruud. 2003. "Contingent Valuation and Lost Passive Use: Damages from the Exxon Valdez Oil Spill." *Environmental and Resource Economics* 25 (3): 257–286.

CEDRE (Centre de Documentation, de Recherche et d'Expérimentation sur les pollutions accidentelles des eaux). 2006. "Report of analyses, oil pollution of the Lebanese shoreline." Report GC.06-21, CEDRE, Brest, France.

Chas-Amil, M. L., E. Nogueira-Moure, M. C. Garcia-Negro, and X. R. Doldán-Garcia. 2004. "Lessons to Be Learned from Past Oil Spills in Galicia (Spain)." *Interdisciplinary Environmental Review* 6 (1): 92–106.

Cohen, M. J. 1995. "Technological Disaster and Natural Resource Damage Assessment: An Evaluation of the Exxon Valdez Oil Spill." *Land Economics* 71 (1): 65–82.

Dimirayak, F., R. Sadek, S. Hraouri-Bloquet, and M. Khalil. 2001. "Marine Turtle Nesting Activity Assessment on the Lebanese Coast—Phase 1: Survey to Identify Nesting Sites and Fishery Interactions." Technical report sponsored by the Ministry of Environment, Republic of Lebanon, Beirut.

DMJM+HARRIS. 2003. "Beirut Suburban Mass Transit Corridor Feasibility Study." Study prepared for the Ministry of Transportation and Public Works, Republic of Lebanon, Beirut.

EC (European Commission). 2006a. "Lebanon 34-Day War: Fact-Finding Mission and Preliminary Damage Assessment." Assessment report, EC, Brussels.

———. 2006b. "Rapid Preliminary Damage Assessment: Beirut and S. Lebanon." Assessment report by the Joint Research Center and European Union Satellite Center, EC, Brussels.

EIU (Economic Intelligence Unit). 2006. "Lebanon Country Report: Outlook in 2007–2008," EIU, October.

Fall, J. A., R. Miraglia, W. Simeone, C. J. Utermohle, and R. J. Wolfe. 2001. "Long-Term Consequences of the Exxon Valdez Oil Spill for Coastal Communities of Southcentral Alaska." Technical Paper 264, Division of Subsistence, Alaska Department of Fish and Game, Juneau.

FAO (Food and Agriculture Organization of the United Nations). 2006. "Damage and Early Recovery Needs Assessment of Agriculture, Fisheries and Forestry." Report TCP/LEB/3101, Special Emergency Programmes Service, Emergency Operations and Rehabilitation Division, FAO, Rome.

GoL (Government of Lebanon). 2006a. "Rebuilding Lebanon Together . . . 100 Days After." Presentation by the prime minister of Lebanon, November 21, 2006.

———. 2006b. "Setting the Stage for Long-Term Reconstruction: The National Early Recovery Process." Presentation at the Conference for Lebanon's Early Recovery, Stockholm, August 31.

Grigalunas, T. A., R. C. Anderson, R. Congar, N. Meade, and P. Sorensen. 1986. "Estimating the Cost of Oil Spills: Lessons from the Amoco Cadiz." *Marine Resource Economics* 2 (3): 239–62.

IUCN (The World Conservation Union). 2006. "IUCN Red List of Threatened Species." IUCN, Gland, Switzerland. http://www.iucnredlist.org.

IUCN/Green Line (The World Conservation Union and Green Line). 2006. "Lebanon Oil Spill Rapid Assessment and Response Mission." Final report prepared by R. Steiner, IUCN/Green Line, n.p. http://www.greenline.org.lb/new/pdf_files/document_2_lebanon_oil_spill_rapid_assessment_and_response_mission.pdf.

Johansson, P.-O. 2006. "On the Definition and Estimation of the Value of a Statistical Life." Working paper 2006–23, Fifth Milan European Economy Workshop, May 26–27. http://www.economia.unimi.it/uploads/wp/Johansson-2006_23.pdf.

Kirch, W., ed. 2008. "Human Capital Approach." In *Encyclopedia of Public Health*, 698. New York: Springer.

Kremer X., A. Pasche, and H. Kilani. 2006. "Palm Island Survey on Saturday, 7th October, and Action Plan for Cleanup Operations." Study prepared for Rempec/Cedre, SWISSAID, and International Union for Conservation of Nature, n.p.

Loureiro M. L., A. Ribas, E. Lopez, and E. Ojea. 2006. "Estimated Costs and Admissible Claims Linked to the Prestige Oil Spill." *Ecological Economics* 59 (2006): 48–63.

MACCSL (Mine Action Coordination Centre South Lebanon). 2007. "Casualties (Civilian & Demining) in South Lebanon from 14 Aug 06 to 03 April 07." Table in *MACCSL Update*, January 12, Tyre, Lebanon. http://lebanon-support.org/resources/UNMACCUpdate12-01-06.pdf.

McCay, D. F., J. J. Rowe, N. Whittier, S. Sankaranarayanan, and D. S. Etkin. 2004. "Estimation of Potential Impacts and Natural Resource Damages of Oil." *Journal of Hazardous Materials* 107 (2004): 11–25.

MOA (Ministry of Agriculture) and UNEP (United Nations Environment Programme). 1992. *Etude de la Diversité Biologique du Liban*. Project GF/6105-92-72. Beirut: Lebanon Ministry of Agriculture; Nairobi: UNEP.

MOE (Ministry of Environment). 2004. "Biodiversity Assessment and Monitoring in the Protected Areas/Lebanon LEB/95/G31: Palm Islands Nature Reserve." Final report for MOE, UNDP, and Lebanese University Faculty of Science, Beirut. http://193.227.177.166/MOEAPP/ProtectedAreas/publications/FinalReport Palm.pdf.

———. 2006a. "The Cost of Oil Spill Cleanup in Lebanon." Internal document, MOE, Beirut.

———. 2006b. "Lebanon's Oil Spill and Its Aftermath." Presentation at the World Bank by Berj Hatjian, director general, MOE, Washington, DC, November 9.

———. 2007a. "Oil Spill Update: February 2007." Online report, MOE, Beirut. http://www.moe.gov.lb.

———. 2007b. "State of the Lebanese Coast after the Oil Spill Caused by the July 2006 War." Press release, May 30.

———. 2007c. "Assessment and Review of Oil Spill Clean Waste and Review of Possible Treatment Options." Preliminary internal report, MOE, Beirut.

Monson, D. H., D. F. Doak, B. E. Ballachey, A. Johnson, and J. L. Bodkin. 2000. "Long-Term Impacts of the Exxon Valdez Oil Spill on Sea Otters, Assessed through Age-Dependent Mortality Patterns." *Proceedings of the National Academy of Sciences of the United States of America* 97 (12): 6562–67.

Murray, J., and A. Lopez. 1996. *The Global Burden of Disease: A Comprehensive Assessment of Mortality and Disability from Diseases, Injuries, and Risk Factors in 1990 and Projected to 2020.* Cambridge, MA: Harvard University Press.

Navrud, S., and R. Ready, ed. 2002. *Valuing Cultural Heritage: Applying Environmental Valuation Techniques to Historical Buildings, Monuments and Artifacts.* Cheltenham, U.K.: Edward Elgar Publishing.

PCM (Presidency of the Council of Ministers). 2007. "Lebanon: On the Road to Reconstruction and Recovery." Periodic report on the post-July 2006 recovery and restoration activities, Second Issue, May 4, PCM, Beirut.

Ramadan-Jaradi, G., 2007. "Effects of Oil on Palm Islands Nature Reserve and Approaches to Long-Term Impact Assessment." Report of the Chair of the Government Appointed Committee of Palm Islands Nature Reserve to the Ministry of Environment, Beirut.

Ramadan-Jaradi, G., and M. Ramadan-Jaradi. 2001. "The Avifauna of Palm Islands Nature Reserve in Lebanon 1893–2000." *Lebanese Science Journal* 2 (1): 17–35.

Sarraf, M., B. Larsen, and M. Owaygen. 2004. "Cost of Environmental Degradation: The Case of Lebanon and Tunisia." Environment Department Paper 97, World Bank, Washington, DC.

SDC (Swiss Agency for Development and Cooperation). 2007. "Cleanup Operations: Palm Islands, Coastline Enfe-Tripoli." Final report, SDC, Lebanon.

http://www.moe.gov.lb/OilSpill/Final%20Report%20Oil%20Spill%20Clean
up%20-%20SDC.pdf.

UNDP (United Nations Development Program). 2006. "Clearance of Rubble and
Debris from Beirut's Southern Suburbs." Project report, UNDP, Beirut.

———. 2007. "Lebanon Rapid Environmental Assessment for Greening Recovery,
Reconstruction and Reform—2006." Report for UNDP, Beirut. http://www
.undp.org.lb/events/docs/DraftReport.pdf.

UNEP (United Nations Environment Program). 2007. "Lebanon Post-Conflict
Environmental Assessment." Report for the Lebanese Ministry of Environment,
UNEP, Nairobi. http://postconflict.unep.ch/publications/ UNEP_Lebanon
.pdf.

Viscusi, W. K., and J. E. Aldy. 2003. "The Value of a Statistical Life: A Critical
Review of Market Estimates throughout the World." *The Journal of Risk and
Uncertainty* 27 (1): 5–76.

Whitehead, J. 1992. "Ex ante Willingness to Pay with Supply and Demand
Uncertainty: Implications for Valuing a Sea Turtle Protection Programme."
Applied Economics 24 (9): 981–988.

———. 1993. "Total Economic Value for Coastal and Marine Wildlife: Specification,
Validity, and Valuation Issues." *Marine Resource Economics* 8 (2): 119–32.

World Bank. 2005. "Islamic Republic of Iran. Cost Assessment of Environmental
Degradation." Sector Note Report 32043-IR, Rural Development, Water and
Environment Department, Middle East and North Africa Region, World
Bank, Washington, DC.

———. 2006. "Lebanon: Economic and Social Impact Assessment from Recovery
to Sustainable Growth. Volume 2." Sectoral Analysis for the Social and
Economic Development Group, Middle East and North Africa Region, World
Bank, Washington, DC.

———. 2007. "Republic of Lebanon: Economic Assessment of Environmental
Degradation due to the July 2006 Hostilities." Report 39787-LB, Sustainable
Development Department, Middle East and North Africa Region, World
Bank, Washington, DC.

CHAPTER 7

Implications for Policies, Investments, and Institutions

Sherif Arif

The work on the Cost of Environmental Degradation (COED) in the Middle East and North Africa region had significant impact on the ground. It has not only raised the countries' awareness about the magnitude of the existing environmental damage, but also driven concrete actions to incorporate the COED into national decision-making processes affecting the environment. This chapter summarizes the region's experiences of the COED's main impacts.

Overview

Since the early 1990s, countries in the Middle East and North Africa have made significant progress in environmental planning and in shaping up their environmental institutional and legal frameworks, assisted by the international community.

National Environmental Action Plans

Countries' adoption of National Environmental Action Plans (NEAPs) marked a turning point in tackling the challenges of managing both the environment and development in the region. The NEAP was the first policy instrument that mobilized governments and international donors in a consolidated effort to address major environmental issues and build environmental capacity. However, NEAPs were not used as instruments for priority setting based on sound cost-benefit analyses; rather, they provided *qualitative* assessments of the state of the environment and natural resources.

Around 2000, when the United Nations Millennium Development Goals were being developed, the World Bank realized that helping countries to prepare or update the NEAPs was not sufficient to ensure environmentally sustainable development in most countries. Furthermore, a rapid assessment of the NEAPs in the Arab Republic of Egypt and Morocco and of NEAP frameworks in Jordan and Lebanon showed that development priorities continued to be affected by the political economy, with disregard to the state of the environment. The NEAPs did not succeed in either demonstrating the economic importance of the environment sector or mainstreaming the environment into the productive sectors of the economy.

Environmental Valuation

The World Bank therefore adopted a different approach with the help of the Mediterranean Environmental Technical Assistance Program (METAP)—a partnership, founded in 1990, between countries of the Mediterranean region and multilateral donors. This approach went beyond the descriptive and qualitative analysis of environmental issues and focused on the economic and financial implications of environmental degradation for the countries' economies (Sarraf 2004, 2005). It catalyzed policy reforms and targeted investments by taking into consideration the cost of inaction or the forgone benefits due to environmental degradation.

As a result, the World Bank's Middle East and North Africa Region took the lead in designing a tool (methodology) to assess the COED at the national and sectoral levels. This assessment process enabled policy makers to internalize the value of the environmental benefits and costs of their social and economic development decisions. Specifically, the COED could serve as an instrument for the following:

• Identify areas where environmental degradation imposes the largest costs to society.
• Identify areas that most significantly undermine social and economic development processes.
• Provide a basis for integrating environmental issues into the financial and economic evaluation of investment projects as well as in sectorwide and economywide policies and regulations.

- Provide a monetary basis for allocation of scarce private and public resources toward environmental protection.
- Enhance the role of the ministries of the environment in demonstrating the importance of environmental protection, using the same "language" as finance and economy ministers.

The COED, undertaken by the World Bank in almost all Middle Eastern and North African countries,[1] has been instrumental in moving the environmental debate beyond the ministries of environment to reach other sectoral ministries, especially the finance ministries. Over the past decades, the COED has had major impacts on decision makers in terms of influencing *national policy dialogue*, increasing *environmental investments*, and strengthening the *capacity of national institutions* in environmental valuation.

The COED's Impact on Policy Making

The COED studies had significant impacts on policy making at the regional, country, and sector levels, as summarized below.

Regional Impacts
At the regional level, the COED results were particularly relevant to the World Bank Environment Strategy, the Euro-Mediterranean cooperation, and the United Nations declarations in the aftermath of the July 2006 hostilities in Lebanon.

Middle East and North Africa Region Environment Strategy. The World Bank Environment Strategy indicated that environmental degradation, poverty, and economic growth are inextricably linked and that environmental degradation exacts an annual estimated cost of 4–8 percent of gross domestic product (GDP) in many developing countries (World Bank 2001a). The World Bank's Middle East and North Africa Region Environment Strategy found that estimating the COED in monetary terms helps to mainstream the environment into the countries' decision-making processes (World Bank 2001b). The region subsequently took the lead in estimating the COED at both the national and the sector levels in most of its countries.

Euro-Mediterranean cooperation. The COED studies have been recognized as a tool to mainstream the environment into other sectors, also at the regional and international levels. At the Second Euro-Mediterranean

Ministerial Conference on the Environment in Athens in 2002, the Ministers' Declaration recognized and welcomed the METAP program's work at the World Bank to evaluate the COED and encouraged future work in the Mediterranean countries (Ministerial Conference 2002). Subsequently, at the Third Euro-Mediterranean Conference in Cairo in 2006, the Ministers' Declaration emphasized that the COED studies carried out under the METAP program demonstrated that environmental inaction is not an option (Ministerial Conference 2006).

United Nations declarations. At the international level, the Report to the Secretary General of the United Nations (UN General Assembly 2007a) and Resolutions 61/194 and 62/188 concerning the "Oil Slick on Lebanese Shores" (UN General Assembly 2007b, 2008) quoted the results of the "Economic Assessment of Environmental Degradation due to the July 2006 Hostilities" in Lebanon (World Bank 2007d). Both the report and the resolutions urged the member states, international and regional organizations, nongovernmental organizations (NGOs), and the private sector to create an Eastern Mediterranean Oil Spill Restoration Fund in support of Lebanon's continuing efforts to address the environmental damage.

Country-Level Impacts
The COED results played an important role in all the Country Assistance Strategy (CAS) studies, several Country Environmental Analyses (CEAs), and national development plans.

Country Assistance Strategy. All the CAS reports and country briefs prepared for Middle Eastern and North African countries by the World Bank have included the COED study results. For example, the COED for Lebanon conservatively estimated the annual damage to the environment at 3.4 percent of GDP in 2000, demonstrating the fragile state of the environment and its importance for the tourism sector and economic growth (Sarraf, Larsen, and Owaygen 2004). Based on those results, the Lebanon CAS for 2005–09 included a pillar on natural resources and the environment (World Bank 2005b).

Similarly, the COED study for Morocco indicated that the cost of water degradation and inadequate potable water, sanitation, and hygiene was the highest environmental damage in the country, estimated at 1.23 percent of GDP in 2000 (World Bank 2003b). This finding supported the Morocco CAS for 2005–09 in considering the improvement of water management and access to water services and sanitation as one of its four key objectives (World Bank 2005c).

Country Environmental Analyses. The COED work also identified priority sectors within the CEA reports developed in Egypt, Jordan, and Tunisia (World Bank 2005a, forthcoming, 2004b). The CEAs used the COED results in the expenditure reviews to analyze the damage and investment costs to mitigate environmental degradation. In addition, the flagship Middle East and North Africa Region Development Report, titled "Making the Most of Scarcity," recognized that environmental problems related to water are difficult to measure but have significant costs that are reducing the current social welfare, as estimated by the annual COED (World Bank 2007c).

National development plans. The COED studies have increased decision makers' awareness of the need to conduct economic assessments to support their countries' policy formulation. For example, in its 10th Five-Year Development Plan, the government of the Syrian Arab Republic required that environmental valuation be included in all environmental policies and programs. The government of the Islamic Republic of Iran has increased its budgetary allocations to its Department of Environment to carry out economic evaluations of the country's ecosystems. The Ministry of Environment in Lebanon has used the results of the COED report as an indicator of environmental performance (Sarraf, Larsen, and Owaygen 2004).

Sector-Level Impacts

Following the successful impacts of the COED estimations at the national level, many countries asked the World Bank to undertake similar exercises at the sectoral level—in particular, the energy, coastal zone, and water sectors.

Energy. The Energy Environment Review (EER) of Egypt was the first report that quantified the environmental damage caused by air pollution resulting from energy use and from burning agricultural residues in Egypt (World Bank 2003a). The country's energy and agricultural burning practices—major sources of air pollution in Cairo—were responsible for the black cloud phenomenon covering Cairo in October and November each year.[2] The EER results enabled the World Bank to advance its policy dialogue in the pollution control sector and to finance the Second Pollution

Abatement Project in Egypt (World Bank 2006a). The results also presented an additional argument that enabled Egypt to ratify the Kyoto Protocol in 2005.

Similarly, the EER for Iran estimated the environmental damage from energy-induced air pollution at 5.1 percent of GDP in 2001 (World Bank 2004a). The country's Department of Environment used the EER results in the preparation of the policy formulation for its 4th Five-Year Development Plan in 2005.

Coastal zone. Given the economic importance of the coastal areas of the Mediterranean countries, many countries bordering the Mediterranean Sea requested the World Bank to carry out COED studies specifically focused on the coastal zones. Through the METAP program, the World Bank financed a subregional study along the coasts of Algiers, Algeria; Alexandria, Egypt (METAP 2006); Nador, Morocco; Northern Lebanon (METAP, forthcoming); and Sousse and Monastir, Tunisia. The key challenge was finding a way to preserve coastal ecosystems threatened by considerable urbanization. These studies enabled policy makers and urban planners to identify the investments needed to protect the most important benefits provided by each coastal zone. Moreover, the Egyptian government is now interested in cofinancing, along with the Global Environment Facility (GEF), an integrated coastal zone management project in Alexandria (World Bank 2009a). Another GEF project is under preparation in Nador, Morocco.

The COED's Impact on Investments

The results of the COED studies were instrumental in increasing budgetary allocations for environmental protection by approximately US$1 billion.

Algeria

Algeria was the first country that used the COED as a tool for priority setting in its National Environmental Action Plan for Environment and Sustainable Development (People's Democratic Republic of Algeria 2002a). As part of the country's first Economic Recovery Program, the government committed US$450 million for the 2001–04 period to implement institutional reforms and pilot investments in the environment sector (People's Democratic Republic of Algeria 2002c). The first investment was geared toward the National Program for

Integrated Municipal Waste Management (People's Democratic Republic of Algeria 2002b).

Egypt

In Egypt, the Second Pollution Abatement Project is the largest environmental project ever cofinanced in Egypt by six institutions—the World Bank, the European Investment Bank, the Japan International Cooperation Agency, Agence Française de Développement, the government of Finland, and the National Bank of Egypt—in the amount equivalent to US$198.4 million (World Bank 2006a).[3] This project focuses primarily on reducing pollution from the energy and industrial sectors to improve air quality. The COED study and the CEA in Egypt highlighted the high damage costs resulting from air pollution, particularly health costs, and played an important role in the design of the project (World Bank 2002, 2005a).

Jordan

In Jordan, the COED study emphasized the environmental damage associated with the Amman landfill, given its proximity to Jordan's capital (METAP 2005). As such, the government, with assistance from the World Bank, prepared the Amman Solid Waste Management Project (US$40.5 million) to strengthen the Greater Amman Municipality's capacity to manage its solid waste, improve the existing landfill, and put in place a Clean Development Mechanism (CDM) project consisting of a landfill gas recovery system (World Bank 2008a).

Morocco

In Morocco, the World Bank (2003b) assessed the damage cost of water resource degradation and inadequate potable water and sanitation at 1.23 percent of GDP in 2000 and the costs of inadequate waste management at 0.5 percent of GDP. The Ministry of Energy, Mining, Water and Environment and the Ministry of Interior used these results to jointly prepare the National Wastewater Program and the National Municipal Solid Waste Program in 2007 (Kingdom of Morocco 2005b, 2007). Subsequently, the Ministry of Finance invested US$100 million in the First Water Sector Development Policy Loan and US$132 million for a Solid Waste Sector Development Policy Loan (World Bank 2007a, 2009c). Furthermore, the government allocated from its own budget about US$25 million in incentives to depollute the Sebou Basin (Kingdom of Morocco 2005a).

Tunisia

In Tunisia, the results of the COED study encouraged the Ministry of Environment and Sustainable Development to increase its investment in solid waste management by benefitting from the CDM to finance improvement in municipal waste management (Sarraf, Larsen, and Owaygen 2004). Through the Sustainable Municipal Solid Waste Management Project, the Ministry of Environment aims to improve the overall management of municipal waste by building new sanitary landfills and rehabilitating polluting dump sites (World Bank 2007b).

The Mediterranean Region

Many Multilateral Development Banks have used the findings of the COED studies to prepare environmental investments in Mediterranean countries. The European Commission cited the results of the COED studies in its communication to the Council and the European Parliament to stress the need for urgent action to safeguard the Mediterranean environment (Commission of the European Communities 2006). This subsequently led to the establishment of the Horizon 2020 Initiative, which aims to tackle the top sources of Mediterranean pollution by the year 2020.

In its first progress report, the European Commission reiterated the importance of the COED studies in improving communication with policy makers and launched a project that would improve assessment and awareness of the societal benefits of environmental improvement and hence political support for environmental policies (Commission of the European Communities 2009). The European Investment Bank financed a study to assess the potential of pollution prevention investments in addressing pollution hot spots in the European Neighbourhood Policy countries[4] and the need for a Mediterranean Hot Spot Investment Program (European Investment Bank 2008). The World Bank, along with GEF, has also launched a new US$50 million regional initiative for an Environmental Mediterranean Sustainable Development Program (Sustainable MED) that includes the preparation of other COED studies (World Bank 2009b).

The COED's Impact on Institutional Strengthening

The COED has served as a tool to enhance the capacity of Middle Eastern and North African countries to estimate the value of environmental cost and incorporate it into policy and decision making. The World Bank organized a series of national workshops, roundtable discussions, and presentations in the region to introduce the COED methodology and discuss the results.[5] Knowledge management and information dissemination were essential to the success of the sessions. The World Bank also prepared a training manual about estimating the COED, which was translated into French and Arabic and was extensively used to support the training sessions and workshops (Bolt, Ruta, and Sarraf 2005).

In addition, the Bank organized the "High-Level Meeting on Economic Tools for Environmental Sustainability," held in July 2003 in Beirut, Lebanon. The meeting presented the COED work already accomplished, provided a forum to exchange the lessons learned, and encouraged regional dialogue about environmental sustainability among the attendees from Middle Eastern, North African, and Gulf countries. Following that meeting, two subregional training courses focused on COED methodology—one dedicated to Mashrek countries (Lebanon, August 2003) and the other to Maghreb countries (Morocco, January 2004). Thereafter, the American University of Beirut developed a course syllabus and complete training materials for academic training in the COED methodologies and application.[6]

Beyond the Middle East and North Africa
Other World Bank Regions have replicated the COED work, mainly by incorporating it into CEAs, as in the following examples:

- In the Africa region, the Senegal CEA includes a comprehensive cost assessment of the health impact of environmental degradation in the country (World Bank 2008b). In Ghana, which faces alarmingly high rates of natural resource and environmental depletion, the CEA estimates the COED at about 10 percent of GDP annually (World Bank 2006b).
- In the Asia region, the Pakistan CEA estimates the annual COED at about 6 percent of GDP (World Bank 2006c).
- In Latin America, the annual COED in Peru is estimated at 3 percent of GDP and the COED in Colombia at 3.7 percent of GDP (Sánchez-Triana, Ahmed, and Awe 2007; World Bank 2007e). The Colombia analysis helped to identify policy options for the second programmatic Development Policy Loan (US$200 million) for sustainable development (World Bank 2007f).

In summary, the COED work performed by the Middle East and North Africa Region has spearheaded innovative thinking in regional policy. It has strengthened policy dialogue and capacity building. It has also increased investment in the environment.

The COED methodology was refined during the past decade. Updating the national-level studies based on this methodology could be beneficial for many countries. Some countries, such as Algeria, Lebanon, Morocco, and Tunisia, have already asked the World Bank to update their COED reports.

Furthermore, there is a need to incorporate the results of the environmental valuation into decision making at the sectoral and national levels, so that the environmental costs and benefits are mainstreamed into national and local planning processes. Particular attention also should be given to building local capacity for understanding the links between poverty and environment, energy and environment, and trade and environment.

Notes

Sherif Arif is a Senior Environment Consultant at the World Bank, 1818 H Street, NW, Washington, DC 20433, USA.

1. METAP, through the Swiss Agency for Development and Cooperation and the Finnish Trust Fund, funded COED studies in Algeria (2002), the Arab Republic of Egypt (2002), Jordan (2005), Lebanon (2004, 2007), Morocco (2003), the Syrian Arab Republic (2004), and Tunisia (2007) (http://www.metap.org). Other COED studies were funded through other sources. For example, the updated COED in Jordan (2010), part of the Jordan Country Environmental Analysis (CEA), was financed through the CEA Trust Fund and the Danish Consultants Trust Fund.

2. The black cloud phenomenon induces low visibility caused by particulates and photochemical smog because of thermal inversion. Pollutants from rice-straw burning and other sources are emitted from the five contiguous governorates of Cairo.

3. The amount is distributed among the World Bank (US$20 million), the European Investment Bank (US$64 million), the Japan International Cooperation Agency (US$52.3 million), Agence Française de Développement (US$59.8 million), the government of Finland (US$1.32 million), and the National Bank of Egypt (US$1.0 million).

4. The European Neighbourhood Policy of the European Commission, developed in 2004, "applies to the European Union's immediate neighbours by land or sea: Algeria, Armenia, Azerbaijan, Belarus, Egypt, Georgia, Israel, Jordan, Lebanon, Libya, Moldova, Morocco, Occupied Palestinian Territory, Syria, Tunisia, and Ukraine." http://ec.europa.eu/world/enp/index_en.htm.

5. The events included those organized in Egypt (December 2002), Kuwait (March 2004), Iran (June 2005), and Tunisia (July 2005).

6. The training materials were used at training workshops such as the one on COED held April 29 through May 3, 2008, in Beirut. http://siteresources. worldbank.org/EXTMETAP/Resources/Cost-of-environmental-degradation-training-manual.pdf.

References

Bolt, K., G. Ruta, and M. Sarraf. 2005. "Estimating the Cost of Environmental Degradation: A Training Manual in English, French, and Arabic." Environment Department Papers, World Bank, Washington, DC.

Commission of the European Communities. 2006. "Communication from the Commission to the Council and the European Parliament: Establishing an Environment Strategy for the Mediterranean." Document SEC(2006) 1082, Commission of the European Communities, Brussels. http://eur-lex .europa.eu/LexUriServ/LexUriServ.do?uri=CELEX:52006DC0475:EN: NOT.

———. 2009. "A Progress Report on the First Three Years of the Horizon 2020." Commission Staff Working Paper SEC(2009)1118, Commission of the European Communities, Brussels. http://ec.europa.eu/environment/interna tional_issues/pdf/med_progress2009.pdf.

European Investment Bank. 2008. "Horizon 2020: Elaboration of a Mediterranean Hot Spot Investment Programme (MeHSIP)." Final report prepared for the European Investment Bank, Luxembourg. http://ec.europa.eu/environment/ enlarg/med/pdf/mehsip_report.pdf.

Kingdom of Morocco. 2005a. "Finance Bill." Kingdom of Morocco, Rabat.

———. 2005b. "National Wastewater Program." Ministère de l'Intérieur/MATEE, Rabat.

———. 2007. "National Municipal Solid Waste Program." Kingdom of Morocco, Rabat.

METAP (Mediterranean Environmental Technical Assistance Program). 2005. "Hashemite Kingdom of Jordan: Cost of Environmental Degradation." Unpublished METAP report, Washington, DC.

———. 2006. "Strengthening of the Capacity in Selected METAP Countries to Assess the Cost of Environmental Degradation in Coastal Areas." METAP Policy Note TF 052354 prepared by researchers of the Finnish Environment Institute, Helsinki, Finland. http://www.accma-maroc.com/docs/docs/ Morocco/Policy_notes_Morocco_en.pdf.

————. Forthcoming. *Legal and Institutional Assessment in Lebanon Coastal Zone and Environmental Degradation, Remedial, and Averted Costs in Northern Lebanon Coastal Zone.* METAP, Washington, DC.

Ministerial Conference. 2002. "Athens Declaration by the Euro-Mediterranean Ministers for the Environment." 2nd Euro-Mediterranean Ministerial Conference on the Environment, July 10, Athens.

————. 2006. "Cairo Declaration of the Euro-Mediterranean Ministerial Conference on the Environment." Third Euro-Mediterranean Ministerial Conference on the Environment, November 20, Cairo. http://www.eeaa.gov .eg/english/reports/events/cairo_declaration_en.pdf.

People's Democratic Republic of Algeria. 2002a. "National Environmental Action Plan for Environment and Sustainable Development (PNAE-DD)." Ministère de l'Aménagement du Territoire et de l'Environnement, Algiers.

————. 2002b. "National Program for Integrated Municipal Waste Management (PROGDEM)." People's Democratic Republic of Algeria, Algiers.

————. 2002c. "Programme de Soutien à la Relance Économique (PSRE): 2001–2004." People's Democratic Republic of Algeria, Algiers.

Sánchez-Triana, E., K. Ahmed, and Y. Awe, eds. 2007. *Environmental Priorities and Poverty Reduction: A Country Environmental Analysis for Colombia.* Directions in Development Series. Washington, DC: World Bank.

Sarraf, M. 2004. "Assessing the Cost of Environmental Degradation in the Middle East and North Africa Region." Environment Strategy Notes 9, World Bank, Washington, DC.

————. 2005. "Cost of Environmental Degradation: Middle East and North Africa Region." Presentation at the World Bank course on COED, Kuwait, June 10–14.

Sarraf, M., B. Larsen, and M. Owaygen. 2004. "Cost of Environmental Degradation: The Case of Lebanon and Tunisia." Environment Department Paper 97, World Bank, Washington, DC.

UN (United Nations) General Assembly. 2007a. "Report to the Secretary General of the United Nations." Sixty-second session, Agenda Item 54, October 24.

————. 2007b. "Oil Slick on Lebanese Shores." Resolution 61/194, Sixty-first session, Agenda Item 53(a), February 6.

————. 2008. "Oil Slick on Lebanese Shores." Resolution 62/188, Sixty-second session, Agenda Item 54, January 31.

World Bank. 2001a. *Making Sustainable Commitments: An Environment Strategy for the World Bank.* Washington, DC: World Bank.

————. 2001b. "Middle East and North Africa." In "Annex A: Regional Strategies" of *Making Sustainable Commitments: An Environment Strategy for the World Bank,* 113–19. Washington, DC: World Bank.

————. 2002. "Arab Republic of Egypt: Cost Assessment of Environmental Degradation." Report 25175-EGT, Rural Development, Water and Environment Department, Middle East and North Africa Region, World Bank, Washington, DC.

————. 2003a. "Egypt: Energy-Environment Review." Reference 8348, World Bank, Washington, DC; Egyptian Environmental Affairs Agency, Cairo.

————. 2003b. "Kingdom of Morocco: Cost Assessment of Environmental Degradation." Report 25992-MOR, Middle East and North Africa Region, World Bank, Washington, DC.

————. 2004a. "Islamic Republic of Iran: Energy Environment Review." Policy Note, Report 29062, World Bank, Washington, DC.

————. 2004b. "Tunisia: Country Environmental Analysis, 1992–2003." Report 25966-TN, World Bank, Washington, DC.

————. 2005a. "Arab Republic of Egypt: Country Environmental Analysis (1992–2002)." Water and Environment Department, Middle East and North Africa Region, World Bank, Washington, DC.

————. 2005b. "Lebanon Country Assistance Strategy." Report 34463-LB, World Bank, Washington, DC.

————. 2005c. "Morocco Country Assistance Strategy." Report 31879-MA, World Bank, Washington, DC.

————. 2006a. "Arab Republic of Egypt. Second Pollution Abatement Project." Project Information Document PO90073, World Bank, Washington, DC.

————. 2006b. "Ghana Country Environment Analysis." Report 36985, World Bank, Washington, DC.

————. 2006c. "Pakistan Strategic Country Environment Assessment." Report 36946, World Bank, Washington, DC.

————. 2007a. "First Water Sector Development Policy Loan." Loan Agreement L 7444-MA, World Bank, Washington, DC.

————. 2007b. "Loan Agreement between Republic of Tunisia and International Bank for Reconstruction and Development (Sustainable Municipal Solid Waste Management Project)." Loan Agreement L7432-TUN, World Bank, Washington, DC.

————. 2007c. *Making the Most of Scarcity: Accountability for Better Water Management Results in the Middle East and North Africa.* MENA Development Report. Washington, DC: World Bank.

————. 2007d. "Republic of Lebanon: Economic Assessment of Environmental Degradation due to the July 2006 Hostilities." Sector Note. Report 39787-LB, Sustainable Development Department, Middle East and North Africa Region, World Bank, Washington, DC.

————. 2007e. "Republic of Peru. Environmental Sustainability: A Key to Poverty Reduction in Peru. Country Environmental Analysis." Report 40190-PE, Environmentally and Socially Sustainable Development Department, Latin America and the Caribbean Region, World Bank, Washington, DC.

————. 2007f. "Second Programmatic Development Policy Loan for Colombia." Program Document, Report 36984-CO, Sustainable Development Department, Colombia and Mexico Country Management Unit, Latin America and Caribbean Region, World Bank, Washington, DC.

————. 2008a. "Guarantee Agreement between the Hashemite Republic of Jordan and International Bank for Reconstruction and Development (Amman Solid Waste Management Project)." L7573-JO, World Bank, Washington, DC.

————. 2008b. "Senegal Country Environment Analysis." Report 48804-SN, World Bank, Washington, DC.

————. 2009a. "Arab Republic of Egypt: Alexandria Coastal Zone Management Project." Project Information Document P095925, World Bank, Washington, DC.

————. 2009b. "The Mediterranean Environmental Sustainable Development Program (Sustainable MED)." Prepared by D. Lotayef for the Sub-Regional Workshop for GEF (Global Environment Facility) Focal Points in the Middle East, North Africa, and West Asia, Cairo, October 20–22, World Bank, Washington, DC.

————. 2009c. "Solid Waste Sector Development Policy Loan." Loan Agreement L7665-MA, World Bank, Washington, DC.

————. Forthcoming. "Hashemite Kingdom of Jordan Country Environmental Analysis." Report 47829-JO, Sustainable Development Sector Department, Middle East and North Africa Region, World Bank, Washington, DC.

Index

References to figures, notes, and tables are indicated by *f*, *n*, and *t* following page numbers. Italic page numbers indicate illustrations.

ECO-AUDIT
Environmental Benefits Statement

The World Bank is committed to preserving endangered forests and natural resources. The Office of the Publisher has chosen to print *The Cost of Environmental Degradation* on recycled paper with 50 percent post-consumer waste, in accordance with the recommended standards for paper usage set by the Green Press Initiative, a nonprofit program supporting publishers in using fiber that is not sourced from endangered forests. For more information, visit www.greenpressinitiative.org.

Saved:
- 14 trees
- 4 million BTU's of total energy
- 1,328 lbs of CO_2 equivalent of greenhouse gases
- 6,397 gallons of waste water
- 1,673 pounds of solid waste

green press
INITIATIVE